The Birds at My Table

The Birds at My Table

Why We Feed Wild Birds and Why It Matters

Darryl Jones

Comstock Publishing Associates
an imprint of
Cornell University Press
Ithaca and London

First published 2018 by Cornell University Press

Printed in the United States of America

Library of Congress Cataloging-in-Publication Data

Names: Jones, Darryl N., author.
Title: The birds at my table : why we feed wild birds and why it matters / Darryl Jones.
Description: Ithaca : Comstock Publishing Associates, an imprint of Cornell University Press, 2018. | Includes bibliographical references and index.
Identifiers: LCCN 2017046553 (print) | LCCN 2017047034 (ebook) | ISBN 9781501710803 (epub/mobi) | ISBN 9781501710797 (pdf) | ISBN 9781501710780 | ISBN 9781501710780 (pbk. ; alk. paper)
Subjects: LCSH: Birds—Feeding and feeds. | Birds—Effect of human beings on. | Human-animal relationships.
Classification: LCC QL676.5 (ebook) | LCC QL676.5 .J66 2018 (print) | DDC 598.072/34—dc23
LC record available at https://lccn.loc.gov/2017046553

Cornell University Press strives to use environmentally responsible suppliers and materials to the fullest extent possible in the publishing of its books. Such materials include vegetable-based, low-VOC inks and acid-free papers that are recycled, totally chlorine-free, or partly composed of nonwood fibers. For further information, visit our website at cornellpress.cornell.edu.

This book is dedicated to Renee Chapman,
Dave Clark, and Josie Galbraith.
Pioneering a new look at an old practice.

CONTENTS

Preface

We love to feed birds. This very day, people throughout the world, from all walks of life, will willingly provide food for wild birds. The foods they offer may vary from discarded food scraps to elaborate home-prepared mixtures or expensively marketed products. These may be simply tossed onto the back lawn or presented via a complex system of tubes and platforms. Such activities may be as casual as a whim or undertaken with a specific goal in mind; they may be part of a vaguely regular routine or a carefully planned strategy. Whatever the process, the central idea is much the same: to provide food for wild creatures, usually close to home. It is often an intimate encounter; we are inviting birds to share our table.

For many of us this can be a profoundly moving experience, an almost magical interaction with nature. Providing food may also be a gesture of care, a heartfelt form of humane assistance to apparently fragile and vulnerable creatures. Lots of people feed birds as a way of aiding their welfare or their preservation, while others simply enjoy seeing wild animals close up. And there are a multitude of other reasons and motivations for feeding, some so obvious that they seem hardly worth mentioning. Feeding birds can be a simple, straightforward pastime as well as something deeply personal we may find difficult to explain.

For many of us, the most important aspect of feeding wild birds seems to be the experience, the improbable opportunity to see and interact closely with real wild animals. This sometimes involves feeding an untamed, untrained, free-flying bird directly by hand. Such an experience is frequently described with words of genuine personal emotion: "privilege, "awe," "moving." More typically, the interaction is a bit more remote yet no less significant: seeing a group of wild creatures heartily partaking of the provisions we have supplied can be genuinely gratifying. These experiences may be different for each person, location, season, and setting. It can change day to day, even hour by hour; that's a key part of the pleasure and delight. Our regular visitors may always be right on schedule, but an unexpected arrival by an altogether unfamiliar species can be an exciting surprise. Keeping at least one eye on the feeder is always a good idea if you have the time; you never know what you might miss!

It's easy to see why this very popular pastime can become an obsession: the pleasure associated with seeing—and providing for—these special visitors each day, with attracting a rare or unexpected species and feeling that you are contributing to the health or continued survival of precious creatures. For some people, these simple pleasures may also become something of an obligation, a daily commitment to uninterrupted provisioning. This dedication—or compulsion—raises some significant questions that also need to be considered. *What would happen to the birds if we couldn't provide these supplies regularly? Would they be able to cope? Would they have to move elsewhere? And might they have become dependent on our handouts?*

This book is an exploration of this fascinating, complex, simple, sometimes compulsive human activity. It is also a serious attempt to understand the reasons why people feed wild birds and a consideration of the possible consequences. One way to start this journey is to visit some bird feeders from around the world as we attempt to unravel this popular, global activity.

It is early morning on November 1, and feeders are getting ready for the day's arrivals . . .

Maine, USA. *The first day of November has brought some unexpectedly snowy weather to the Tilleys' well-treed garden in the hills of northern*

Maine. Up at dawn as always, Janice Tilley is surprised by the dusting of new snow that greets her, but she is well prepared. "Time for my special gorp," she announces proudly, retrieving several string bags of her own concoction (made from a well-guarded family recipe) from the battered fridge in the garden shed. "All I will tell you is that I add seven ingredients to the lard, including walnuts and grape jelly. And the nuthatches love it!" Winter is always tough for the birds that don't escape the cold. "The fat really seems to help; they scoff at the sweet bits, but it's the lard that builds them up. I honestly think that my winter provisions are keeping a whole bunch of chickadees alive till spring."

Gwynedd, Wales. *For Jim Griffith, who lives on "the windy side" of the Snowdonia Mountains in northern Wales, filling his collection of homemade "little house" feeders is now a daily preoccupation. The wintry winds have seen off many of his fair-weather visitors, but he still has a solemn duty in providing for some diminutive and secretive residents, a pair of robins, as well as some special new arrivals. "They might look delicate, but these redbreasts—they're tough little blighters," says Jim, who has watched them taking turns to feast on the mealworms and peanut cake he replenishes every day at 6 a.m. and 2:30 p.m. "My robins live right here," he says, "but the siskins seem to come in from the woods just for the thistle seeds." Today he waits, broom in hand, by his back door, to ensure that his favored clients are not disturbed by the unwelcome and "downright brazen" jays. "The robins are my most reliable friends these days," Jim states quietly. "I don't know what I would do without them."*

Wellington, New Zealand. *Early November weather is typically unpredictable in Wellington, at the bottom of the North Island of New Zealand, even though it is supposed to be summer. "Sunny with a bloody good chance of rain," jokes Susie McGan, though her frustration is evident. The frequent showers are seriously disrupting her attempts to draw in the Tuis, one of the native sweet-toothed species that flock to her simple sugar-water feeder. There are always plenty of Waxeyes [Silvereyes] and the ubiquitous House Sparrows, but it's the Tui that Susie really enjoys. "So full of life and energy, they always fill me with joy." Here, providing regular seed attracts a remarkable variety of introduced species such as Goldfinches, turtle-doves and sparrows. "I love them all, I really do," she explains, "It's just that these Tui were really rare so recently. It's such a blessing to have them coming to visit now. By offering a little something* *that they appear to enjoy, I seem to be offering hope—hope that we might be able to restore some of the damage we humans have done. Well, the return of the Tuis makes it seem possible."*

Brisbane, Australia. *November in humid Queensland typically means occasional wild, thundery summer storms with brief periods of ferocious rain, followed almost immediately by blinding sunshine. It's a backdrop that seems entirely appropriate to the spectacular primary-colored plumage of the most conspicuous visitors to my humble seed platform. Rainbow Lorikeets seem improbably exotic to have become the most abundant bird in most of Australia's larger cities. Loud, excitable, and extroverted, pairs of these gaudy parrots now dominate my feeding station, although they also tend to feast and fly fairly quickly. This allows the less pushy species—rosellas, magpies, doves, and honeyeaters—a chance to taste what remains. All these birds are common and widespread, clearly not needing assistance because of climate or conservation. "Feeding is folly," according to my otherwise friendly neighbor, Richard. "All you are doing is encouraging the dictatorial!" he regularly admonishes. "All these species are dominators; the smaller birds we are trying to attract by plantings don't have a chance with these bullies around!" He has a point, and he is not alone in expressing these forthright views. Why do I feed? It's got something to do with connecting with nature, I think, but I'll need to give this some careful thought.*

These vignettes offer a small but informative glimpse into the complexities of what is usually regarded as a simple, often strictly private, pastime. Each of these situations is typical, even commonplace, yet they also introduce some of the personal—and sometimes universal—issues associated with feeding. What to provide as food and when to provide it? Should the food change with the seasons? What to do about unwelcome visitors (including the greedy [insert local villain species] that chase our favorites away? Is my feeding disrupting the natural balance, or is it replacing something that has been lost? And, really, why do we do this anyway?

Feeding wild birds is very popular. A number of surveys from the United States, the United Kingdom, Germany, New Zealand, and Australia consistently report that sizable proportions (often about half) of households are engaged in some form of feeding.[1] That's a lot of people providing a lot of bird food. This food may be table scraps, picnic leftovers, homemade products such as suet balls, or commercial seed mixtures. The most regular form of feeding takes place at home, in gardens, yards, or balconies, usually with the food offered in locations that allow the birds to be viewed clearly. In addition, a popular form of wild bird feeding occurs in public spaces such as urban parks and recreational areas. This form of feeding is usually less organized than feeding birds at home, typically involving the casual tossing of picnic leftovers to the gathered gulls and waterbirds. In contrast to such impromptu feeding, however, is the virtually universal pastime of "feeding the ducks down at the lake," a practice that leads to untold tons of bread being tossed to waterfowl the world over, bringing joy to millions and often resulting in the eutrophication of urban lakes and a host of attendant ecological problems.

Bird feeding has also become a massive industry on a global scale. After humble beginnings as a largely DYI domestic activity, the production of commercial products such as seed mixes and the associated hardware has become a gigantic multinational enterprise. In the last few decades the amount of food being provided has reached new highs. In the United States it is estimated that about 60 million people supply hundreds of thousand tons of seed to wild birds every year.[2] In the United Kingdom the amount of seed sold would support many times the actual population of the birds being fed.[3] Globally, well over a million tons of seed are sold as wild bird food each year, much of it grown in India and Africa for export to the rest of the world. The United Nations estimates the global bird-seed industry to be worth US$5–6 billion, growing around 4% annually since the 1980s.[4]

Astonishingly, most of this vast amount of bird food is being consumed. In almost all circumstances, the food being eaten by the birds is, by definition, extra or supplementary to their normal diet. Our provision of what is *additional* food is a subsidy to what they would be able to glean during their natural foraging activities. While this may seem a bit obvious, it is worth pointing out explicitly: the food we offer wild birds is entirely on top of what they would find themselves. It is the influence of this additional food in the lives of the birds we feed that we need to understand.

The role of food and nutrition is of fundamental importance to all animals, especially in relation to the success or otherwise of breeding. Because of this, a large number of experiments have investigated the effects and influences of adding extra foods to the diets of a wide range of wild animals, including reptiles, mammals, and many species of birds. These supplementary feeding studies (covered in detail in Chapter 5) demonstrate conclusively that even a little additional food can have significant consequences. These might include having a higher chance of surviving the winter, laying more and slightly larger eggs, and raising extra offspring. For almost all species, it is fairly safe to predict that more food is likely to mean more individuals in the future. That may be exactly what we hoped for, especially for the species we care about. But how many is too many? What might happen to the local community of all bird species when the abundance of those taking advantage of our feeding grows but not the rest? And what about the unwanted birds, the aggressive and dominant species, or those that we may regard as pests?

The types of food we provide may also be of great significance. Today, the nutritional qualities and standards of most commercial seeds are carefully monitored and maintained. Nonetheless, it is all too easy to find poor quality mixes, as well as spoiled, tainted, and even poisoned bird foods. More fundamentally, almost none of the types of foods we offer to birds are a natural component of their diet. Birds the world over adore black-oil sunflower seeds, for instance, the single most popular seed sold internationally, yet very few of the species flocking to our feeders—or any of their ancestors—will have ever fed on sunflowers in the wild. At least these are mainly granivores, adapted to consuming the seeds and grains available in their local environment. For just about everything else that is offered to (and gleefully accepted by) our wild visitors—bread and food scraps, chips and fries, sausage and cheese—will certainly be very different from anything in their natural diets. Even the most expensive commercial seed mix may not provide the same balance of proteins, fats, and vitamins they would have obtained from their natural diet.

But does this matter? If the birds are loving whatever it is we are providing, isn't that enough? Wouldn't they avoid inappropriate foods? A moment's reflection from the human perspective should be sufficient to suggest that what we eat really does matter. We are surrounded by the physiological consequences of people eating too much salt, sugar, and fat even when we know we shouldn't. Furthermore, the nutrition we provide in our feeders may only be "supplementary" to all the other things they eat during the day, or it may be all that is available. And this raises one of the thorniest issues of all: To what extent are the birds we feed actually dependent on us? Has our desire to be close to the birds, to share our table with them, in fact shackled us together as providers and recipients? Or do most birds simply treat our provisioning as a trivial part of their overall diet while they continue to forage naturally?

These are all critically important questions, and yet, despite the massive scale of the activity and the millions of people actively and passionately involved, remarkably little is known about most such issues. A lot of excellent and relevant research has been undertaken, though many of these studies are quite specific to particular places and species. Part of the mission of this book is to unearth and explain this research.

Exploring the reasons that lead us to spend our money on expensive seeds and feeding gadgets, to invest enormous amounts of time and energy to provide the best feeders, to battle daily with squirrels, crows, and possums, all in the hope of attracting some wild birds, is fascinating, frustrating, and fun. It is also vitally important to confront some of the uncomfortable questions raised here. Because it really is about what is best for the birds.

I'd like to close this preface with one note on terminology. Ever since I started the process of putting this book together, the problem of "feeder" has confounded me. The word describes both the people engaged in feeding birds as well as the apparatus used to provide the seed. I have tried various alternatives, but these were just too cumbersome or artificial. I'm afraid we are stuck with the one word throughout, so please be aware of the context.

Acknowledgments

Many books claim to be on some sort of journey, either in a metaphorical sense or in relation to actual geographical travel. Many are a bit of both, traversing intellectual landscapes as well as real places, complete with unexpected discoveries, life-changing experiences, faulty guidance, dead ends, and long stretches of apparently featureless terrain. The journey may be exhilarating and fundamentally rewarding, but it is always good to arrive somewhere, even if it is nothing like what was envisioned at the start.

The journey of this book was no different. I set out confidently on what I naively thought would be a prolonged but well signposted journey, in a straightforward manner toward an indistinct but fairly certain destination. Such are the first steps of many a pilgrim. The reality was far less direct, much riskier, and infinitely more interesting. And as is so often the case, the arrival at the end turned out to be barely the beginning.

Writing that is based on the published works of others—the standard approach of anyone attempting to synthesize a research topic—can be deceptively simple, particularly now that virtually everything is instantly available online. Read carefully with a critical eye, note the key findings, and relate the conclusions to your accumulating body of knowledge. That is supposedly how it works. Except as an active researcher myself, I am acutely aware that the nice, tidy findings presented in the colorless prose of the typical scientific journal article are usually only part of the story. So often it's the things that went wrong or turned in unexpected directions—the stuff that does not get past the journal editors and peer reviewers—that can provide some of the most important insights.

With this in mind, I decided to contact almost everyone still alive who had made a contribution to the scientific study of garden bird feeding to ask whether I could discuss their work and maybe just talk about the activity informally. Even though this is not a particularly large field, I did not really expect to get much of a response. For some, the research had been undertaken some time ago, while others had retired, moved on, or were now pursuing other interests. All true, yet almost everyone I successfully tracked down said they would be happy to "talk." Indeed, it soon became obvious that the topic of bird feeding was still of passionate interest to these people. The positive responses resulted in a multitude of e-mail exchanges, Skype conversations, and even old-fashioned telephone calls to locations all over the world, frequently at odd and inconvenient times of the day. These interactions were at least as important as the published science in the evolution of the ideas and perspectives that eventually became this book.

But these many conversations, mainly via electronic technology, were just part of the complex pattern of information that was being woven together. Over about a year I was fortunate enough to meet face to face with a remarkable number of the researchers I previously knew only as authors of studies I had read. (Yes, there was a real and extensive geographical journey.) As I had anticipated, it was the chance to hear, in person, the raw details and realities behind some of the key research and stories from many of the most significant scientists and personalities involved in wild bird feeding that proved to be invaluable. Hearing their insights and personal views would simply have been impossible without these meetings. I remain astounded at the generosity and honesty of so many people, and their willingness to share their thoughts with a virtual stranger.

I am deeply and sincerely grateful for every one of those delightful and inspiring encounters and will long treasure them. A long amble around the damp streets of Cardiff with Richard Cowie; a magical day with Josie Galbraith, among the Hihi and Takahe on Tiritiri Matangi Island; an illuminating and extraordinarily candid visit with Chris Whittles in Shrewsbury; the revolutionary day when I visited both the Royal Society for the Protection of Birds (RSPB) and the British Trust for Ornithology (BTO) in England, where a simple query about feeding birds in summer with Will Peach, Kate Risely, and Kate Plummer led to an entire unplanned chapter; watching the snow build up with some trepidation while talking for hours with David Bonter and Emma Greig at the Cornell Lab of Ornithology; ticking off a whole page of new birds with Piotr Tryjanowski in the fields close to his home near Poznan in Poland. And then there were equally open and informative meetings with Mel Orros in Reading; Ralph Powesland in the Marlborough Sounds; Andre Dhont, Steve Emlen and Natalia Demong in Ithaca, New York; Jim Reynolds in Birmingham and London; Hannah Peck in Penryn; Arienne Prestor in Malmo; Mark Cocker in Caxton; Monika Rhodes and Ann Göth in Sydney; and Saren Starbridge, Tim Low, and Rich Fuller in Brisbane. I thank them all sincerely.

Similarly, I am deeply indebted to the people I met via Skype or telephone: Sharon Birks (USA), Stephen Schoech (USA), Dan Chamberlain (UK), Dave Clark (UK), and Peter Berthold (Germany).

No less important, and often equally detailed and passionate, were e-mail exchanges (sometimes in addition to other interactions) with Grischa Perino and Monika Rhodes (Germany); Neil Gladner (USA); Abel Julien (Spain); Leah Burns (Iceland); Ralph Powesland and Eric Spurr (New Zealand); Will Peach, Pat Thompson, Jon Blount, and Emma Rosenfeld (England); Alexa and David Rase (South Africa); and Julie Lake and Brendan Trappe (Western Australia).

The writing of this book has been remarkably global in its development. The first words were composed in Queenstown, New Zealand (watching Pukeko being fed bread beside an almost frozen Lake Tekapo), the last at a writer's retreat at Mount Tamborine in southern Queensland, Australia. In between, significant sections were composed in Copenhagen, Denmark; Malmo, Sweden; Canberra, Australia; Singapore; Maliau Basin, Borneo, Malaysia; and the British Library, London. By far the greater amount of writing (and related thinking), however, occurred in the delightfully productive quiet of the State Library in Brisbane, Australia.

This book had a relatively long gestation and I am immensely grateful to the following colleagues for their long-term support and encouragement: Mike Toms from the British Trust for Ornithology, Jim Reynolds from the University of Birmingham, Tim Birkhead from the University of Sheffield, Mark Cocker from Caxton, Suffolk, Rich Fuller from the University of Queensland, and Tim Low from Brisbane.

For permission to undertake the travel associated with the development of this book, I am extremely grateful to Professor Hamish McCullum, the head of the Griffith School of Environment, and Professor Zhihong Xu, the director of the Environmental Futures Research Institute, both at Griffith University in Brisbane, Australia.

During the concluding periods of writing I benefited enormously from two invaluable periods of writing retreat. The first was a thoroughly informal arrangement as the guest of Ralph Powesland (previously of the New Zealand Department of Conservation) at his remote and spectacular retirement home in the Marlborough Sounds of New Zealand (where the local Weka have successfully trained Ralph to supplement their diet every day, though he seems to imagine it is some sort of experiment designed by him!).

The second retreat was as a recipient of a BREW (Bush Retreats for Eco-Writers) environmental writer's fellowship in the inaugural residency at the latest location, the subtropical rainforests of Mount Tamborine in the southern mountains of Queensland. It would be difficult to imagine a setting more conducive to writing and thinking. I am extremely obliged to BREW and especially to Sandra Sewell for her sensitive hospitality.

Of course, none of this would have been remotely possible without the solid, reliable foundation of the home front. Somehow, Liz, Dylan, Caelyn, and Manon all successfully feigned interest in my endless stories about feeding and shared my excitement at seeing a new bird at the feeder. They may even have put out the seed on occasions.

But without any doubt, my greatest thanks are extended to three extraordinary friends and colleagues who have shared the frustration, excitement, tedium, and wonder of attempting to undertake research in the field of bird feeding: Renee Chapman, Dave Clark, and Josie Galbraith. This book is much more informative and was infinitely more fun to write because of their multifarious and invaluable contributions: ideas, research discoveries, field trips, anecdotes, rare insights, and forgotten documents, as well as cakes, pints, jokes, beds, and encouragement.

Finally, I offer my sincere thanks to my agent Margaret Gee for her support and guidance. And at Cornell University Press, Kitty Liu, Susan Specter, and Meagan Dermody have been an author's dream to work with through those inevitable tedious bits at the end.

The Birds at My Table

1

Why Bird Feeding Matters

The birds at my table are impatient. If I am late, they line up along the veranda railing to peer accusingly at me though the kitchen window. Do I dare make my coffee before fetching the seed container? If my delay is particularly prolonged, they may start to yell, as though encouraging (threatening?) me into action. When I open the door to step out and gently deposit the usual handful of "wild bird mix" on the circular feeding platform, they continue to grumble, staying just out of reach, until I retreat and close the glass door quietly behind me. As soon as I am inside, they quickly fly to the platform to start feeding, deftly husking the sunflower seeds but always keeping an eye out for potential disturbances: a sudden movement inside the house, the ever-present cat, the arrival of potential competitors in the nearby foliage.

Although they appear tame and are thoroughly unperturbed by the presence of people, the birds at my table are truly wild creatures. They are a native species living as they always have done—finding mates, raising offspring, flying freely, and foraging widely—largely unnoticed by the humans they share their habitat with. Although these birds live their lives among us in the strange landscape of the suburban ecosystem, their complex and fascinating family lives and interactions are effectively invisible. Unless we are able to entice them out of the trees and into our lives.

By providing something they are always searching for—food—we may be able to bring such wild birds into view. If they respond, their visits are a privileged and sometimes intimate view into their world. They don't need to come. But they do. They accept my invitation and fly in, to spend a little of their busy day close to the humans, but on their own terms. If they don't like the situation, if they sense possible danger or see something unfamiliar, they are gone, in a flash, without apology or explanation. And there is nothing I can do about it.

While they are present, however, I have the opportunity to watch and maybe even learn something. On one level, I simply take in their beauty and personality. Most of our impressions of free-flying birds are fleeting and fragmentary, a mere glimpse of a restless shadow. When this relentless energy actually halts, abruptly, at the feeding station so close by, the colors and delicate patterns of the feathers are arresting. The plumage of even the most commonplace species, no matter how familiar from a distance, is indeed marvelous when viewed closely. But the most important impression is of a living being, an individual pausing briefly to refuel (or just snack) before carrying on with its own day. We can never really know but it's hard not to wonder: Where are they going in such a hurry? Who will they meet? How do they decide? Do they have priorities? Yes, of course such thinking is a form of anthropomorphism, the attributing of human traits such as decision making and planning to animals. But as we learn more about the behavior of animals, the traditional distinctions that have long separated us from them are steadily being challenged. I can never pretend to know what is going on behind these bright avian eyes, but as our gazes meet (over the seeds and coffee cup respectively), and I start to mentally arrange my day ahead, it's pretty easy to imagine that the birds are doing the same.

And these are thoughts resulting directly from my act of feeding. By providing something that attracts these birds to my yard and then the feeding station on my balcony, I have brought wild birds into the private and personal space of my home. In the process, through a moderate level of quiet observation and respect, these animals have changed gradually from general examples of their species to known individuals, with distinct personalities and features. They haven't changed (as far as I can tell), but I have: my life now includes important elements of fascination with and concern for these creatures. My house is in, after all, a fairly typical suburb, full of threats and opportunities for wild visitors, all associated with our human-dominated mode of living. We (people generally) chose to clear the original natural landscape for places to build houses; we selected the garden plants and introduced new predators (such as this deceptively lazy cat), among many other alterations. All these typical features of people-space affect the wildlife living around us. Where will they live and breed? Will they find enough food? How does my humble feeding table fit into their lives? Indeed, is the food I provide actually useful, or suitable, or maybe even harmful? These were questions I had not even considered before I started feeding.

That is because I have come to realize that the birds at my table are also *guests*. And that makes me their host with the attendant responsibilities of anyone inviting friends to share a meal. A much larger table inside the house (but within easy view of the bird feeder) has supported many lunches and dinners for human visitors. Around this table, the ancient conventions of hospitality and neighborly communion are accompanied by generous provisioning and attention to personal preferences. (Though, if we are being honest, it has been my wife who has been the primary provider and arbiter of taste.) The unspoken objective is to have satisfied visitors who are able to enjoy the food and company and who eventually leave content and healthy. The "sharing of bread" is a universal and important cross-cultural means of social interaction and trust building. I would obviously be horrified if these guests rejected the food or became ill as a result of eating it. All necessary care would have been exercised to prevent such outcomes; such are the responsibilities of a host.

As the host to birds, similar expectations apply. I am responsible for the quality and quantity of the "meal" and the hygiene of the "plate." These guests tend to be fairly flexible with respect to what is eaten and clearly have their favorites and dislikes; some things are consumed instantly, others ignored completely. They are, however, decidedly lax about their table manners and often leave the table in a less than hygienic state, seemingly unconcerned about what my next guests might think. Cleaning up after the party is also part of the host's charter.

Almost none of these issues occurred to me way back when I initially decided to see what might happen if I placed some seed on the feeding station. This platform, built onto the balcony railing outside the dining room, had been constructed by the previous owners of the house. According to our neighbors, however, feeding had not been practiced here for many years. Would birds come back after such a long time? A handful of sunflower seeds started a vague experiment, but the ongoing results have been unexpectedly significant, both personally and professionally. The first birds arrived a few days later and have continued to do so, more or less daily, ever since. That is hardly big news, but this simple activity—providing food for birds—has generated a seemingly endless sequence of questions and concerns. Given that this is also an activity practiced by millions of people, these issues might similarly be considered around the world. What began so simplistically for me has led to profound experiences and unexpected dilemmas, productive dialogues as well as bitter disagreements. From a handful of seed to a topic of global importance. You see, I am convinced that bird feeding really matters. In this book I will try to show why.

My Table Is in Australia

The birds at my table are probably rather different from those at yours. Regardless, I suspect that many of the issues, concerns, challenges, and rewards are likely to be similar for feeders wherever they live. The details will differ, but this is a topic of international interest. Feeding is, however, global though not universal; although people all over the planet feed wild birds, there are also places where this does not occur. Similarly, there are places where feeding is encouraged and actively promoted, and others where it is banned or at least frowned upon. Even among birders, conservationists, and environmental activists, opinions can be sharply divided. Such differences in attitudes and sometimes policy are fascinating, and worth exploring. Why such contrasts? What are they based on?

Such conflicting views have been particularly strong in Australia.[1] As we will see, while Australians are subject to some of the most strident anti-feeding attitudes, statements, and policies, they remain among the most active and passionate feeders of wild birds globally. This is much more than the celebrated antipodean "antiauthority" attitude so often cited by tabloid commentators. For large proportions of the population to pointedly defy the widespread "Do Not Feed" messages suggests something significant is going on. This is just one of the reasons (there are others) why our exploration of international wild bird feeding starts amid the suburban backyards of Australia.

I live in Brisbane, Australia, a city of about 2 million people situated on the coast about halfway up the eastern side of the continent. It is a rapidly growing region in a lush, subtropical landscape. Although suburbs now sprawl in all directions, a combination of sensible planning, simple geography, and blind good luck has left lots of areas of native eucalyptus forest among the roads and houses. These bushland patches, many quite large, support a remarkable diversity of wildlife species, including birds, a number of which leave their natural habitats to explore the nearby suburbs. Our human habitats are far from bare and unwelcoming to native birds. The warm, humid climate allows almost anything planted to grow quickly, leading to suburbs of bewildering varieties of trees and shrubs.

Like most gardeners everywhere, Australians love showy flowers, and although traditional European varieties such as roses and gladiolas are still common, over the last few decades many species of Australian native trees and shrubs have become popular as suburban plantings. Brisbane gardens are full of shrubs and small trees such as grevilleas, callistemons, and banksias, all natives although they often originate from distant parts of this large country. The flowers they produce are gorgeous, large, and conspicuous, and they seem to come in all shades of red, yellow, orange, and even purple. To the ecologist, these colors immediately announce their adaptive objective: to attract birds. And not just any birds. While the color may act as a neon "Open for Business" sign, it is the abundant supply of nectar deep within the base of the flowers that the birds are attracted to. These sweet rewards are typically located deep inside the blooms, requiring specialized equipment to gain access. This is often in the form of the long beak and the even longer brush-tipped tongue found in one of the most characteristic bird groups of the Australian avifauna, the aptly named honeyeaters.[2] This huge family (the Meliphagidae) consists of more than 90 small to moderately sized birds occupying every possible habitat throughout the vast and diverse Australian continent, provided there are nectar-bearing plants present. In terms of both the diversity and density of such plants, however, no natural landscape has anything like the nectar supplies of the suburbs.

Enhancing the availability of native plants is one of the fundamental objectives of "wildlife gardening" where attracting birds is usually the predominant goal. If you have the space, trying to replicate components of the bird's natural environment is an obvious platform for providing habitat and enticing wild visitors. Habitat always starts with vegetation, though not all plants are suitable for attracting birds. For example, it is now well established that native species of birds make better use of indigenous plants than they do of species originating elsewhere. For the majority of birds, this relates to the availability of insects they can use as food—native plants support more invertebrates than do introduced species—and native plants are also more likely to be used for nesting and perching sites as well. In general, flowers may offer nutritional resources, such as nectar and pollen, but far more important for birds is what pollinated flowers result in: fruits and seeds. These concentrated bundles of energy and protein are crucial sources of bird food and are key elements in a well-designed wildlife garden.

The general principles of wildlife gardening are all about creating habitat, hopefully for as many species as possible. In the case of the Australian nectar-producing plants we are discussing here, however, we seem to have a problem. Popular shrubs such as grevilleas and callistemons, often promoted as "bird magnets," provide a massive sugar supply, and, thanks to clever selective breeding for prolonged display, one that is unnaturally long-lived and reliable. Add to that the abundance of these plants and you have a suburban landscape literally dripping in super-sweet, bird-attracting native plants. Surely that must be a good thing for urban bird conservation in Australia.

That is certainly what was expected, and so-called bird-scaping gardens through the planting of bird-attracting plants has been encouraged by Australian bird and conservation organizations for decades.[3] Indeed, such garden design is often strongly promoted as an alternative to feeding: "Bring them in with nectar, not seed." It was an activity that worked only too well, but not at all as hoped. Certainly, the plants did attract large numbers of birds seeking the nectar, but instead of increasing the diversity of birds to apparently "enhanced" suburban gardens, all too frequently the result was lots of one species: the Noisy Miner. A typical native honeyeater found throughout the eucalyptus bushland of eastern Australia, this bird should be welcome in our gardens, or so you'd think. It is most certainly not, for the simple reason that if you have Noisy Miners, you don't get anything else.[4] Noisy Miners live in complex and tight-knit social groups that aggressively defend their territories. Territorial activity is normal behavior for the majority of songbirds during the early breeding season. Noisy Miners, however, take this to the extreme by remaining savagely despotic all year round, and most significantly, driving away all other species from their defending area. The only species able to withstand these relentless attacks are birds bigger and, in some cases, even nastier than the miners. This unfortunate but otherwise natural behavior by a common native bird species is now recognized as a major "threatening process" in Australia, seriously undermining the conservation status of numerous smaller species. The threat is most acute in the woodlands outside the cities but has been catastrophic for people living in the suburbs trying to bring in the birds. The nectar-rich garden has turned out to be a honey trap.[5]

Overrun by honeyeaters? It seems to have little relevance to wild bird feeding, doesn't it? Not if you are a typical feeder doing your best in the suburbs of much of Australia. No matter what you try, the presence of miners often means not much else in the way of smaller species. What started as seemingly sensible wildlife gardening decisions has often resulted in a disastrously depleted bird community. This is a cautionary tale of unintended consequences stemming directly from our attempt to provide a key resource for the purpose of attracting birds. Planting particular types of shrubs may appear to be more "natural" than putting out seed, but the effects on the birds may be just as pronounced. Providing any sort of food for wild creatures is an experiment with a range of possible outcomes. One of the simple messages of this book is that changing the foods that birds can eat almost always changes other things as well.

For a large part of Australia, the wholesale planting of nectar-bearing vegetation has indeed changed the community of birds in the suburbs. I have witnessed the dramatic consequences among the varieties of birds that are able to visit my garden and eventually my feeder. Earlier I suggested that the birds at my table were probably different from those at yours. There are several reasons for this, beyond the simple fact that you may live in a different country. First, as mentioned, I live with a colony of

Noisy Miners. Not a lot, but this colony of about a dozen effectively keeps all the smaller species away, almost completely. Imagine your garden or feeders without any of the smaller birds: no tits or chickadees, no hummingbirds or Silvereyes, no finches or robins. I hope you see my point.

Thankfully, I live within a well-treed suburb and not far from a large conservation reserve. As a result, I am fortunate to have visits from a wide variety of the slightly larger—and miner-resistant—species, including several species of parrots such as cautious Pale-headed Rosella, raucous Sulphur-crested Cockatoos, feisty butcherbirds, and jittery but animated Spangled Drongos. These birds are the more occasional visitors, always welcome but, being somewhat casual, they are much harder to get to know. No, when I consider the real birds at my table, the reason for my efforts and concern—and the genuine impetus for this book—it comes down to just two species, Rainbow Lorikeets and Australian Magpies.

The Birds at *My* Table

Rainbow Lorikeets are simply outrageous. They are outlandishly colorful, ridiculously unaware of risk, and disarmingly confident. No bird so conspicuous and extroverted should have the audacity to also be the most abundant bird in a majority of the large cities of Australia. But this is a status recently confirmed; numerous surveys of Australia's urban birds have found that these spectacular birds now outnumber all others.[6] This includes the usual global dominants such as feral pigeons, House Sparrows, and blackbirds—all of which are well established in Australia—as well as the ubiquitous corvids (crows and related species). The proliferation of lorikeets is a relatively recent phenomenon; traditionally, lorikeets were more likely found in open eucalyptus woodlands characteristic of the Australian bush. Why have so many become city dwellers? As is so often the case, the answer seems tied to food.

Lorikeets are a fairly distinctive group within the broader parrot group, and are distinguished from their relatives by their remarkable tongues. The business end of this specialized apparatus is covered in a dense forest of soft, fleshy bristles, giving them the rather obvious name of "brush tongues." While lorikeets do eat seeds like regular parrots, the structure of their tongues indicates that they are also adapted to exploiting a particular resource: the nectar and pollen produced by the predominant trees of Australia, the eucalyptus. As with so many ecological cycles in this unpredictable land, the flowering of these trees is highly erratic, but when conditions are right, vast landscapes may be transformed into a sweet-smelling banquet. As expected, a great diversity of Australian animals—beetles and moths, marsupial gliders and possums, bats and many different birds—quickly take advantage of these bonanzas.[7] From the big ecological perspective, this is all about ensuring pollination, of course, but the lorikeets don't care. With their specialized feeding equipment and belligerent personalities, they dive right in, determined to make the most of the all-too-brief boom times. When it's all over, they are back to gleaning whatever they can find.

Unless they are among the many lorikeets now living in the suburbs. Not only are these birds able to take advantage of the same nectar-bearing shrubs as the Noisy Miners (and easily win any interspecies arguments with the much smaller honeyeaters), they also benefit from the types of larger trees found in domestic gardens and city parks and along the streets. Because of the usual human preference for large, colorful, and prominent flowering displays, the species and varieties selected for these places are typically showy. Such trees are also often developed selectively for prolonged displays. While these preferences resulted in a range of spectacular tree species from distant lands (such as the scarlet flowers of the African tulip tree and the gorgeous purple jacaranda from South America), the same criteria have been applied to the selection of eucalyptus trees for ornamental planting. As mentioned previously, interest in using native Australian plants for gardens has revolutionized the way homeowners and local planners have designed their plantings. But while the trees chosen may indeed be native, they are often species (as well as cultivars bred to enhance their display) with larger and longer-lasting blooms than just about anywhere on the continent. It's what people want to see. And what we appreciate aesthetically, our brush-tongued nectar-feeding visitors enjoy as an unnaturally prolonged supply of their favorite foraging resources. Instead of an irregular, temporary, and unpredictable boom, the wide diversity of eucalyptus now available across the Australian suburban landscape ensures that the busts are brief, if at all. It is hardly a surprise, then, to find that these clever, adaptable, and opportunistic birds have prospered spectacularly.[8]

The abundance of their main natural diet across the local landscape is apparently the chief reason that my most regular bird-table visitors are Rainbow Lorikeets. They almost certainly don't need the seeds I diligently provide, and their periodic absences are probably due to some massive flowering by a eucalyptus clump nearby. I guess this puts my feeder into perspective. And as much as I enjoy their presence and vibrancy, I can't pretend that we are really that close. Why, I can't even be sure that they are the same birds each time!

There are no such qualms when it comes to my other special guests. Compared with the brash party-animal presence of the lorikeets, the arrival of the magpies is quiet, discreet, and professional. Their demeanor and presentation could hardly be more different; these birds are strictly black and white. You know where you stand: you're equals, and yet it can also be personal. OK, I will stick my neck out and say it: these are birds you can develop a relationship with. Really.

First, though, some relevant background. The Australian Magpie may have pied (black and white plumage) coloration, but it is not closely related to the various magpie species of the Northern Hemisphere. These "true" magpies are corvids and part of the large group of intelligent birds that includes crows, jays, and ravens. The confusingly named Australian Magpie is really a "butcherbird," a group of mainly carnivorous birds found in all habitats but especially the open woodlands. Australian Magpies spend much of their day wandering around on the ground foraging for worms and grubs just below the surface. Their feeding behavior means that the best substrates for magpies to seek their prey in are where the soil is soft and moist;[9] watered and regularly mown suburban lawns are therefore ideal, and magpies have become one of the most successful and familiar suburban species throughout Australia.[10] Again, this is due to human actions influencing the availability of a key food.

The seed I place on my feeding table is not exactly what Australian Magpies regard as a key food group, but they come anyway. They do consume seed regularly, but it is not their main reason for visiting. To gain the attention of birds like magpies, quite a different diet is required. So periodically my bird table is scattered with a variety of foods most people would be astonished, if not horrified, to see on a feeder: meat. If you seek such a fish, however, you need the right bait—and for a big carnivore such as a magpie, the most successful lures are distinctly nonvegetarian.

Magpies will eat almost anything, but typical feeder items are ground beef (minced beef), sausage, salami, and canned or dry dog food, as well as meaty leftovers from the family's last meal.[11] Of course, none of these are in any way "natural," and many feeders are legitimately concerned about whether such foods may be suitable or harmful (this topic is explored in Chapter 6). What would be far more appropriate would be real live worms and grubs, but a ready supply of these would be difficult to obtain. Instead, I try to provide that standard live insect food, mealworms, whenever possible. Mostly, though, it's raw mince or dog food.

The magpies visit my bird table only after the lorikeets have left. Usually, they wait quietly in a nearby tree, flying over quickly should they see that meat is present (the lorikeets having ignored this, of course, but strange things can happen . . . see Chapter 9). Because magpies are strictly and permanently territorial, I know that *these* birds are the same individuals each time. The male, distinguished by the clear, clean white-and-black nape, always arrives first, his mate soon after, but only if there are no other birds of any species anywhere nearby. The female (her neck color graduating from white to black) is far more reticent than her mate, and spends most of her time at the table scanning for trouble. In contrast, the male magpie is exceptionally trusting (I am avoiding saying "tame"), often continuing to feed when I carefully and slowly approach. And just occasionally, he will approach me, especially if I hold a piece of meat or cheese (or he thinks I do). Interacting with a wild creature such as this, often accompanied by a gentle caroling, can be genuinely moving. Who could resist that bright quizzical eye, the slanted head, the deft plucking of the morsel from my fingers? I didn't train him, but I suspect this bird knows all about training me. He gets his treat, but for me, it's a treasured experience.

As the magpie skips away a little to consume the item, I—again—consider some of the many elements to this minor interaction between human and wild bird. The sheer possibility of hand-feeding a wild animal is the end point of a long and complex series of manipulations of natural situations. Even if we ignore the landscape-scale changes associated with urbanization (loss of habitat, replacement of the original vegetation with plants selected for nonecological reasons), the species of wildlife now living among us in urban environments are very likely to share several important characteristics. These will include fundamentals such as overcoming their natural fear of people, a tolerance of human disturbances, and a willingness to try new things.[12] Shy, easily disturbed, risk-averse species are unlikely to do well in human-dominated environments. Bold, assertive, inquisitive, risk-taking animals, in contrast, are much more likely to be the ones making it in towns and cities. While there are plenty of obvious exceptions, these are traits typical of the birds that visit our feeders. A far more extreme group make up the very few that are confident enough to take food from our hands.

And Then a Few Questions Come to Mind

Stepping back a little, simply visiting a feeder marks a decidedly "unnatural" behavior for most birds. Whatever structure we use—horizontal platform, vertical tube, covered hanging bird house, ceramic bell, elaborate squirrel-proof multientrance "avian apartment tower"—it will not be anything like what our birds will have encountered during their entire evolutionary history. Yet somewhere, some bird sometime had to overcome its strong and sensible fear of a new thing in its world before it landed, and its companions and descendants were able to follow suit and reap the benefits. Even more fundamentally, these pioneering birds had to recognize that the items being offered were actually food. It is worth remembering that most of the food routinely provided on feeders throughout the world is not a traditional part of the diet of the species consuming it. While the birds obviously love certain types of feeder foods (think black sunflowers, suet, even bread), can we be certain that these items are nutritionally suitable or even harmless? Just because a food type attracts plenty of takers, does that mean it should be used?

If I wander down this particular route in my considerations (the magpies have withdrawn because a crow has arrived), it's hard to escape the most fundamental of all these questions: *Should I be feeding wild birds at all?* This is a question rarely heard in many parts of the world but is almost unavoidable in Australia where opponents of feeding are ubiquitous and vocal. Everyone—feeders included—knows the standard arguments: feeding leads to dependence; it spreads disease; it increases aggression; it leads to poor nutrition; it disrupts movements (and probably causes global warming as well). These potentially negative impacts are acknowledged around the world, but they are definitely the dominant narrative in Australia. Any possible benefits are typically dismissed or belittled, if mentioned at all.[13] As a result, the practice of wild bird feeding here is much more private, a discreet activity rarely acknowledged or discussed. The antifeeding stance of almost all bird and conservation organizations also means that Australian feeders find it difficult to obtain advice or hints on how best to feed. We all know that the answer to "How?" will be "Don't!"

The attitude almost everywhere else in the world could hardly be more different. As evidenced by the public statements made by most of the major bird and conservation groups of the Northern Hemisphere, it is the apparent benefits of feeding that are clearly prominent if not self-evident. Again, the reasoning will be fairly familiar: feeding assists birds through the winter, enhances breeding and survival, supports many species of conservation concern, and engenders interest in environmental issues.[14] There are many more benefits named along similar lines. My guess, however, would be that the majority of people engaged in wild bird feeding (outside Australia, at least) have relatively little exposure to the various arguments and debates concerning feeding. Perhaps these are all a bit academic and remote. After all, feeders are participants in a very popular pastime that seems to have a clear message (simply put): "If you care about birds, feed them!" Furthermore, advice, hints, tips, and commandments are everywhere, all promoting feeding and advocating ever more associated hardware, squirrel guards, species-specific seed mixes and opportunities to share your experiences and even contribute to scientific understanding.[15] No wonder the feeding industry in Europe and North America continues to expand. A largely private activity has become a very big business on the strength of personal experiences and community participation. "Feeding is popular with the people; it must be good for the birds" seems to be the reasoning.[16]

It would, however, be both mischievous and unfair to suggest that the messages from the Northern Hemisphere are all positive and that the numerous concerns are being ignored. This is most certainly not the case. All the prominent groups provide clear advice on key issues such as minimizing the spread of disease and situating feeders to reduce the chance of predation. Indeed, discounting the periodic hysterical antifeeding articles and sensationalist exposé, many of these issues are regularly discussed and debated; there are certainly plenty of dissenting opinions about all manner of topics.[17] The sheer scale of interest and participation in wild bird feeding provides a lively platform for the exchange of ideas, allowing any interested feeder an endless supply of suggestions, advice, and warnings.

I would make the claim, however, that some of the most important questions concerning feeding have not been adequately answered and some have not even been asked. Indeed, the truly astonishing thing is that for such a common and popular activity, participated in daily by millions of people worldwide, supplied by a huge and effective industry, and ably promoted by prominent advocacy groups, there remain many aspects of feeding we really do not understand. The amount of reliable research into most of these issues has been remarkably limited. It is fair to say that many of both the strident claims of the opponents ("Feeding causes dependency" or "Feeding leads to poor chick nutrition") and the equally confident statements of the proponents ("Feeding increases chick production" or "Feeders are better conservationists") have not—until very recently—been based on sound scientific studies.

And this really matters. Like any other person providing food to wild birds, I would be alarmed if I was to learn that this simple pastime was in any way detrimental to the very creatures I care for and about. Similarly, I would be pleased to learn that my actions were actually benefiting these wonderful animals. But this is most decidedly not simply about just me—or just you. Each of us might be concerned mainly about our own backyard, but we need to be aware that we are also part of a truly international activity, a colossal experiment on a global scale.[18] That's going to be a large amount of supplementary food and a very large number of participating birds. I think that it is time to take a close and careful look at what we are involved in.

A Supplementary Feeding Experiment on a Global Scale

I have made some claims as to the scale of wild bird feeding already, with bold statements about the numbers of people involved and the amounts of foods provided. Almost everything written about this activity includes some version of such figures, usually gleaned from the same relatively small number of studies. While this is fairly normal practice in science writing, it is also important to return to the original publications to make sure we are being accurate. This approach is an essential part of the approach being taken in this book: if the story that unfolds is to be accepted, every aspect has to be balanced and fair yet skeptical and critical. I fully expect my own ideas and explanations to be challenged. So let's start with the numbers of feeders we are talking about.

Before we start, we need to clarify what this actually means. The figures provided by researchers are almost always in the form of a proportion (or percentage) of a group. Stating, for example, that "157 people surveyed in a particular place said that they provided food for birds" is meaningful only if we know how many people made up the whole sample and how these people were selected or contacted. The resulting proportion is usually called the "participation rate."

The earliest and very widely quoted estimate comes from one of the first quantitative investigations of the patterns of bird feeding in suburban gardens. The study was conducted by British researchers Richard Cowie and Shelley Hinsley in the 1980s.[19] They reported an unexpectedly high figure of 75% of households engaged in bird feeding. Although the study was conducted entirely within two suburbs of northern Cardiff in Wales, this figure has been endlessly cited as being representative of the United Kingdom as a whole. It was, nonetheless, a level of participation that surprised everyone, including the researchers. Indeed, despite many subsequent surveys from several countries, the Cardiff figure remains among the highest reported anywhere. For this reason alone, it is entirely reasonable to ask why, because it suggests potential problems with many similar surveys.

First, it is important to note that Cowie and Hinsley were not attempting to come up with a definitive proportion of the population of their city who were engaged in feeding birds (and they were certainly not expecting their data to be extrapolated to the entire country). Rather, they were much more interested in what the feeders (the people) were actually doing: their stated aim for this study was "to assess the types and amounts of food that are provided and how it was presented." Their work was pioneering and remains influential, but it was designed primarily to obtain information about the activities associated with bird feeding. Their data were collected the hard way, by door-to-door delivery of a questionnaire, with the responses collected in person a few days later. Even with so direct and personal an approach, the response rate will never be perfect; people will be away (or hiding), mistakes may be made when completing the forms, and some people will simply not be interested. Such was the case in Cardiff: of the 287 hand-delivered surveys, 73% (209) were able to be counted in the study. The response rate was somewhat higher than most mail-out surveys, and the researchers would have been relatively pleased with their efforts. Nonetheless, this meant that more than a quarter of the sample were not completed for unknown reasons or, more important, because the respondents expressed "no interest." The unresponsive component of the survey were subsequently assumed by the researchers to be nonfeeders. This leaves us with the 157 surveys mentioned above, which provides the now famous 75% (157 of the 209 acceptable completed surveys).

This rather tedious divergence into primary-school arithmetic does have a point. Rather than being an indication of "the proportion of Cardiff households engaged in feeding" (or even of the UK), the figure cited would more appropriately be called "the proportion of respondents of two samples of suburban Cardiff who returned surveys." The people who responded were also more likely to complete the questionnaire because they were feeders, and maybe because they were pleased someone had asked about their pastime. These are normal problems associated with the completion of questionnaires on any topic, and there are various ways that researchers can deal with them. My point is that a common form of bias is likely to affect many similar surveys: they are likely to result in an inflated proportion of feeders responding than nonfeeders. The figures are still useful, but we must not assume that they are truly representative of the community being sampled.

With that caution in mind, what else did Cowie and Hinsley find? It turns out that the 75% refers quite specifically to the proportion of Cardiff feeders who fed during the winter months. Throughout the warmer months (May to September), the level of participation dropped to less than 40% of households (among the survey respondents). But even allowing for generous overestimation, these figures still represent a lot of people. For those interested in the new and relatively unexplored fields of the ecology of towns and cities—"urban ecology"—such a high level of influence over foraging resources was definitely worth noting. At the time, however, only a few scientists were taking urban ecology seriously.

A Brief Detour via Wagga Wagga

Around this time (the early 1980s), I was a young and naive student attempting to understand something about the rapidly changing bird community of my hometown in the dry interior of New South Wales. I had grown up in the quintessentially Australian town of Wagga Wagga (pronounced "Wogga Wogga," Wiradjuri Aboriginal for "many crows"), a thriving country center famous for producing well above the expected number of sporting stars. Being demonstrably useless at most sports, I followed my growing interest in natural history to university elsewhere but returned home to undertake my first independent research project. The decision to study the birds in town rather than in natural areas away from the influence of humans was not viewed particularly favorably by my peers and academic advisers. In terms I was to become all too familiar with during the following years, these critics regarded true ecology as being about entirely natural environments, devoid of any elements of human activity. If ecological research required avoiding all anthropogenic influences, urban environments were simply impossibly tainted. For many traditional ecologists, the very concept of "urban ecology" was just too hard to accept.

These were theoretical considerations for the future, however, and with the strong support of other mentors I plunged into the wilds of Wagga. The impetus for my interest in local urban birds was the recent and unexpected arrival of a new bird: the European (Eurasian or Common) Blackbird. While the town supported a large and diverse suite of native species (including crows, of course), there were also plenty of largely British species that had been introduced during the disastrous period of "acclimatization" in the 1860s.[20] These introductions included House Sparrows and Tree Sparrows, Greenfinch and Goldfinch, and Starlings. The addition of the Blackbird to this list in this location was, however, completely unanticipated. We all knew this jaunty songster from the cooler, wetter, more English-weather places in southern Australia, but to see them scampering about among the dry-country garden plants of the arid inland seemed all wrong. How could they survive here? How did they get here? What were they finding to eat? These were the types of questions that initiated my lifelong fascination with wildlife in cities. And also with the way the human residents interact with these creatures.

Although I was vaguely aware that some people around town regularly fed birds in their gardens, it was during my bird surveys that I slowly began to realize that the concentrations of magpies, parrots, and native pigeons I was recording related to several generous feeders. Quite clearly, the presence of feeding stations was having a major influence on the density of certain species, both positively (for those attracted to the food) and negatively (for those wanting to avoid these aggregations). This early contact with bird feeders also exposed me to the sometimes strong attitudes feeders held toward certain species or categories of feeder visitor. For example, magpies were always welcome, corvids never; colorful plumage was favored over black or dull; and finches (even including the introduced species) were especially enjoyed while the sparrows barely tolerated. These attitudes and opinions were simply part of the entertaining but hardly relevant background to my "real" quantitative data on species richness and diversity. I would realize only later that such information was just as important to my attempts at understanding bird feeding.

Estimating How Many People Feed Birds

The Wagga Wagga urban bird study[21] was the first of a long series of such projects conducted over the next few decades, which I continued when I moved to complete my graduate degree at the newly established Griffith University in Brisbane. Through a combination of strategic career planning, professional inertia, and satisfaction with the optimal location, I have now been based in Brisbane for over thirty years. Throughout this time I have been involved in a wide variety of urban ecology research projects involving many different approaches and disciplines. The central objective of almost everything I have engaged in has been to attempt to understand more about the way people interact with wildlife in urban settings. This broad canvas has included attempts to solve wildlife-human conflicts, base wildlife management plans on behavioral insights, and discern why some species prosper in cities while others flounder. Key to all these projects has been a clear appreciation that understanding any of the inherent complexity associated with people-animal-ecosystem interactions is beyond a single researcher, methodology, or approach. Any success we have had comes down to a willingness to listen and learn and remain open to insights from unexpected places.

The foundation of this diverse suite of interrelated studies started with the fortuitous gathering of a group of keen but unsuspecting graduate students in the late 1990s and early 2000s. Somehow attracted to the general theme of wildlife-human interactions, each student pursued quite different projects, but all eventually included elements of relevance to bird feeding. Sometimes such aspects were unintended last-minute additions to a larger study. For example, Leoni Thomas's multifaceted investigation of the origins and management of the many wildlife conflicts in the local region included a detailed questionnaire.[22] Such survey instruments require careful development and testing to ensure that all the questions make sense to the respondents. Toward the end of this process, we were discussing the wording of a section aimed at finding out about people's gardening practices. Out of the blue, someone casually asked, "What about bird feeding?" It is a measure of how far we have advanced to have to admit that we had not even considered this before. Feeding was simply not regarded as significant; the prevailing attitude was along the lines of: "Everyone in Australia knows you shouldn't feed, so it's going to be trivial." Andrew Cannon, then at the British Trust for Ornithology, had already articulated this attitude: "Generally the more conservation-minded and knowledgeable individuals in Australia do not feed their garden birds."[23] We agreed, and so did almost everyone we spoke to. I recall stating confidently at the time, "Less than 10% feed, I'm sure," when asked to predict the outcome.

Leoni's questionnaire was mailed out to addresses in Brisbane picked randomly from voter rolls because we really wanted a true and representative sample of the general community. Although the eventual results were to provide invaluable information on a range of wildlife management issues, it was the responses to the straightforward question on feeding ("Do you provide food for wild birds at your home?") that subsequently received all the attention. Contrary to all expectations, Leoni's data indicated that 38% of respondents across the suburbs of Brisbane (Australia's third-largest city) were engaged in feeding wild birds. This was big, unexpected, and for a lot of people unwelcome news. In presentations to community, conservation, and bird groups given soon after, we were often confronted with a range of reactions, from outrage at the "stupidity and ignorance" of these benighted people to that of genuine surprise at finding that a phenomenon so common could also be so invisible. Most frequently, however, was outright denial: it can't possibly be true; the data collection, the methods, the analyses—somewhere we must have made a serious mistake. The latter reaction was so prevalent that we too began to doubt the findings.

And then the results of other studies began to roll in. Directly in response to Leoni's Brisbane results, Briony McLees conducted a similar questionnaire survey in the southern city of Melbourne.[24] I suspect she was motivated at least in part to show that Victorians were different from Queenslanders. They were: even more of them fed birds, with 57% of respondents admitting to the activity. Back in Brisbane, Peter Howard, investigating wildlife-human conflicts using various social science methodologies, and Dan Rollinson, who was studying the urban ecology of Australian Magpies (see Chapter 5 for further details), both undertook specific surveys of the feeding practices and attitudes of people in the Brisbane region. Despite taking entirely different methodological approaches, the apparent participation rates they discovered were almost identical: 38% and 37%.[25] In 2007, Go Ishigame and Greg Baxter conducted a hand-delivered questionnaire survey (à la Cardiff) in rural and outer suburban areas in western Brisbane and obtained participation rates of between 36% and 48%.[26] Admittedly, having so many samples from the same geographical area is not ideal, but this much data does point to an unequivocal conclusion: at least a third of the people in the Brisbane area feed birds. We are still guessing and extrapolating wildly, however, if we were to extend these figures to the whole of Australia. If only we had some sort of national sample.

The goal of gaining a broader picture of feeding across the country became a real possibility when I was approached by the national public media organization, the Australian Broadcasting Corporation (ABC), to assist in developing an exciting urban wildlife project. *WildWatch*[27] was something entirely new for Australia in that it was an attempt to discover how people from all over the country interacted with wildlife around their homes. The innovative aspect of this ambitious project was its use of a variety of media to both engage and communicate with the public. It all started with a high-quality television documentary that explored the extent to which wildlife now lived among us in our towns and cities. This show directly challenged the perception that wildlife lived away from the big towns, "in the bush" as Australians like to say, beyond the city limits. The documentary concluded with an invitation to participate in a Web-based survey, including an opportunity for people to post their own wildlife stories. Participation in the *WildWatch* survey was also continuously encouraged by regular features on ABC radio shows, an effective network of radio stations that form the only truly coast-to-coast media coverage in the country. Finally, on the basis of public responses to the survey, two follow-up television documentaries were produced, featuring real stories of people and the species they cared about. I was enthusiastically involved in this process as an adviser and particularly in the development of the online survey. Researchers are notorious for wanting to add too many questions, but I could hardly let the opportunity pass and successfully argued for a couple of questions on feeding. Well, that was the plan. Its success depended entirely on whether people were willing to be involved. In fact, could we be sure that Australians cared enough about the wildlife in their backyards to bother?

We need not have worried. An extraordinary 27,364 people completed the survey, one of the largest samples for an entirely voluntary survey ever conducted in Australia. Now, clearly, there were going to be important biases with this sample: the demographic of ABC viewers and listeners tends to be somewhat older, better educated, and possibly more computer-literate than the Australian norm. Even more important, people already interested in wildlife are much more likely to respond. Pleasingly, however, the proportion of respondents from each of the states was very close to their respective segment of the national population. On the other hand, though not unexpected given the emphasis on urban wildlife, more respondents were from cities than from rural communities. So, accepting all those provisos, what did the *WildWatch* survey tell us about feeding for (slightly older, city-dwelling, public broadcaster–supporting) Australians? A total of 28% fed birds with seed and 29% hand-fed or fed on the ground. And this was across the country: no state or territory had less than a quarter of respondents as feeders.

Enough on Australia. This exercise emphasizes the lack of even the most basic data relating to feeding activity. We are about to explore what is known from the United Kingdom and the United States, but apart from the handful of studies from Australia mentioned here, the only other quantitative information on participation rates are from New Zealand.

This may simply be due to a reliance on English-language publications, of course, though I have attempted to search far and wide for relevant news. I suspect that wild bird feeding—at home at least—occurs in most places in the developed world to some extent, and is far less likely in developing regions. But these are guesses at best; we simply do not have the data. If wild bird feeding was somehow related to economic level or local community activities, is it possibly being practiced in the newly affluent suburbs of Brazil, China, and South Africa, for instance? I will examine these issues later. For the moment, however, we have to accept that our international perspective is extremely limited.

In New Zealand, for example, scientific interest in bird feeding is unusually pronounced, perhaps because of its use in supporting several species of endangered birds. This topic is explored in some detail in a later chapter because researchers have learned some critical lessons concerning the impact of supplementary feeding. As is well known, New Zealand is home to some of the most vulnerable populations of birds anywhere in the world, the result of historic habitat destruction and predation by introduced predators. As a result, this small, remote country has developed innovative and successful approaches to conservation, including the use of additional foods. Far less interest, however, has been shown in wild bird feeding gardens. In a land where awareness of the plight of so many native species is high, the fact that most of the birds encountered in backyards are typically English imports suggests that these species are coping well enough without help. Nonetheless, feeding is common, though no one really knew much about this until recently.

The New Zealand Garden Bird Survey[28] single-handedly changed that deficiency. Almost entirely the work of a single person—the indefatigable Eric Spurr—and strongly supported by the government science agency Landcare Research Manaaki Whenua, this citizen science project was initiated in 2007 and has run annually ever since. The work is closely modeled on similar private garden–based surveys such as the much older Garden BirdWatch[29] in the UK. The New Zealand version runs on a more modest scale with around 3000 participants from throughout the country sending in information on the birds they see at home. As with the other garden bird surveys, observers provide invaluable data on trends in populations and insights into what are largely urban birds that would be simply impossible to obtain by other means. In addition to questions about birds, respondents are also asked about their pets and gardening practices and, of course, whether they feed birds. The national participation average over the seven annual surveys up to the latest in 2014 has remained very close to 70%. This is certainly a comparatively high participation rate but not that unexpected in light of the unavoidable sources of bias already discussed.

Building on Eric Spurr's general findings for New Zealand, in 2013 researcher Josie Galbraith conducted some of the most detailed studies of feeding attempted anywhere.[30] Josie employed a carefully designed questionnaire that was mailed to randomly selected households in the six largest cities spread throughout the two main islands of New Zealand. Given its relevance to this book, we will be delving into the findings of this important research many times. At the moment, however, we will report simply the participation rate determined by Josie: 63% of all respondents reported feeding birds, though her response rate was only 26%. Again, while this figure has probably been boosted by the greater likelihood that feeders would respond more often than nonfeeders, it seems clear that over half the households in New Zealand are feeding birds.

Several recent and important studies from the United Kingdom and the United States respectively are worth mentioning here because they, along with the New Zealand work, were focused investigations of feeding at quite specific locations—and are hopefully evidence of a new level of interest in these issues. The first study was conducted in the city of Reading in southern England by Melanie Orros and Mark Fellowes.[31] Among many significant findings based on interviews with over 500 people, they obtained a figure of 55% of households feeding birds. A similar examination of the English city of Leeds found feeding rates varying from 38% to 70% across 18 neighborhoods.[32]

In surprisingly rare US work on this topic, Christopher Lepczyk and coresearchers investigated regional differences in feeding practices among households in Michigan and Arizona.[33] They found important differences between the two areas, with an average of 66% of Michigan residents feeding birds compared to 43% in Arizona.

All the surveys mentioned so far have been constrained and biased in some ways, all understandably and almost inevitably given the small scale of the studies and the typically modest levels of funding involved. While all provide important indications of the extent of participation in bird feeding for the places surveyed, it would be silly to extrapolate these findings too far. For reliable statistics on feeding activities at national scales, enormous resources would be required, something more appropriate at the level of governments. Thankfully—and rather unexpectedly—such national surveys do exist for the United States and the United Kingdom, providing us with the best picture of feeding at the scale of an entire country. In both cases, the feeding component was a minor part of much broader investigations. Nonetheless, the inclusion of this topic among the aims of the research suggests some level of formal recognition of this pastime within the two countries involved.

Since 1955, the Federal US Fish and Wildlife Service has undertaken regular surveys to gauge levels of public involvement in a range of outdoor recreational activities across the nation. The National Survey of Fishing, Hunting, and Wildlife-Associated Recreation provides data on trends and details such as the amount of money spent and whether the activity occurred at home or away, and has useful state-by-state comparisons.[34] The 2011 edition was the twelfth and is currently the most recent of these surveys to be released, with the one before it appearing in 2006. The results are gold for researchers as well as those involved in wildlife management, and each new survey is awaited with great anticipation. We can certainly include researchers interested in bird feeding among this group.

Following methodology established for the 1991 survey, the 2011 version was conducted by the US Census Bureau and involved personal interviews with an astounding 48,600 households from all states, followed by smaller samples from particular target groups. This laborious process ensures that "everyone" (meaning samples from all social, economic, and racial groups) is included, not just those that actually fish, hunt, or feed. For our present purposes, this really matters. Wildlife feeding is included in the section covering "Observing, Feeding, and Photographing Wildlife" and is reported separately for those participating at home or away. For example, in 2011 the total number of participants engaged in these general activities was 72 million, although fully 95% of them did so at home.[35] Of these, an astounding total of 53 million people reported feeding wildlife, almost exclusively birds. These figures equate to a national participation rate in bird feeding of 74% for the United States. Given the sound methodology used, this is probably the most reliable rate so far reported.

The national picture for the United Kingdom is similarly derived from large-scale, government agency–instigated surveys—three in this case, though each had rather different objectives. The UK Department of the Environment, Food and Rural Affairs published a report in 2002 entitled *Working with the Grain of Nature,*[36] a proposal for addressing a variety of environmental problems through increased community engagement in activities such as wildlife gardening. Among the comprehensive material presented was the estimation that 60% of British households with a garden feed wild birds, a practice identified by the UK government as a key indicator of the health of urban biodiversity.

The second UK survey—CityForm[37]—was a large multidisciplinary investigation of economic, social, and environmental sustainability of urban environments. This major questionnaire-based study sampled households from five large cities across Britain (notably excluding London): Oxford, Leicester, Sheffield, Glasgow, and Edinburgh. Among the fifty questions asked were several on wildlife gardening; again, possible bias was minimized by embedding the feeding questions within a wide range of queries about other issues. When all the data were considered as a whole (the cities were not compared separately), the CityForm survey found that 53% of participants fed birds. Note that all respondents were from major urban centers and not from smaller towns or rural areas, and that the response rate of returned surveys was 30% to 40%.

Finally, the largest UK survey of relevance here is the annual Survey of English Housing, a government-sponsored study of general household characteristics, activities, and attitudes.[38] Like the US Fishing and Hunting Survey, it is based on face-to-face interviews, in this case, around 20,000 respondents chosen randomly from across the country. The feeding data, from the 2001/2002 edition, which also included some wildlife gardening questions, yielded a figure of 64%.

Let's bring this discussion on feeding participation to a simple conclusion. In the end, we have only a handful of studies that have published estimates of the proportion of feeders among the communities studied, and these vary considerably in scale and reliability. I would place the two large-scale interview-based surveys—the US Fishing and Hunting and the UK English Housing surveys—in a "best estimates" category, for both the sound nature of their methods and their truly national coverage.[39] This suggests that the figures 74% for the United States and 63% for the United Kingdom are probably close to the actual situation. The other studies mentioned offer important insights into feeding participation over more limited areas, although the figures—ranging from 37% to 75%—are likely to be somewhat higher than reality. What is far more important than these lifeless percentages, however, is the staggering scale of feeding they represent: going on the most recent data, that's around 53 million households in the United States and 13 million in the United Kingdom who spend their money to provide food for wild birds. That's going to amount to a lot of cash. Indeed, recent estimates indicate that US households spent something like $4 billion on bird food and an additional $10 million on associated equipment such as feeders and nest boxes.[40] In the UK the best estimates of seed sales are $440 million and a further $220 million for continental Europe.[41] (The evolution of this specialist industry will be explored in the following chapter.)

Of course, what most of this money is being spent on is bird food—mountains of it. In the United States, feeders distribute 450 million kilograms (990 billion pounds) of seed each year, while Britons offer 60 million kilograms (132 million pounds) to their avian visitors.[42] And this is just the commercial products. Bread, for example, is still among the most widely used food for birds. Josie Galbraith found this item to be the main type of food distributed to garden birds in New Zealand, with a total of over 5 million loaves being consumed.[43] One sobering, even astonishing, statistic, calculated by Gillian Robb and her colleagues, is that the amount of seed distributed in the United Kingdom each year could sustain a theoretical population of 30 million Great Tits, one of the most abundant garden species.[44] This is six times the actual population of the species! In a similar vein, Melanie Orros and Mark Fellowes calculated the energy component of the foods supplied by feeders in Reading.[45] This pioneering study accounted for all food types provided, which included bread, kitchen scraps, and suet, as well as all the usual seeds. Their meticulous analyses found that the average Reading household provided 628 kilocalories of energy per day, and that this, extrapolated to the entire country, was capable of supporting about 71 million birds of the ten commonest feeder species. And that, again, was around twice the estimated national population size of these species. As an interesting (but very relevant) sideline, these researchers also calculated that if the local gray squirrels pinched only one-tenth of the food, it would be sufficient to support a minimum of a million of these animals, even if they ate nothing else.

If most of this food is actually being consumed by wild birds (leaving aside the squirrel quotient), doesn't that simply mean that they need it, that our feeding is supporting local populations? Or does more food lead to more birds, either attracted to the feeder in the short term or eventually, by an increase in the number of offspring produced or perhaps because of better survival over winter? Indeed, what *is* the effect of this additional food on the birds we are feeding? And what about the other so-called nontarget species that inevitably take advantage of the foods in supply? Be they grackles, feral pigeons, or jays—let alone rats and raccoons—feeders almost always attract some species that are simply not welcome. And predators, too, like Sparrowhawks and Sharp-shinned Hawks, as well as the neighbor's cat?

If we are honest and attentive, it doesn't take long to come up with a set of potentially unsettling questions. For an innocent, feel-good pastime, a private and satisfying activity that appears to do plenty of good, surely these concerns are unfounded? Don't we know enough already?

So What's the Problem?

For most feeders, it's fairly easy to carry on and enjoy your hobby, content in the knowledge that the birds seem to be happy, the food is being eaten, and only last Thursday you saw your first (nominate latest "table tick'" species). These are among the privileges of unofficial membership in the international wild bird feeders club, a worldwide collective of disparate individuals focused primarily on their own backyard. It's just that now we know that an awful lot of backyards together represent a very large change in a fundamental component of any habitat: the diversity and amount of food. As well as enjoying the many benefits associated with feeding birds, feeders also need to accept that this practice is highly likely to change things.

We are probably all aware of the main general concerns associated with feeding. We may even have read articles and blogs that discuss these issues and the debates and discussions that follow.[46] Do birds become dependent? Does feeding spread disease? Should I feed during the warmer months? These questions, and many others, are important and require properly reasoned responses. There are endless personal opinions out there, but we need the advice and pronouncements to be sound and scientific. While much remains to be investigated, enormous advances have been made recently in our understanding of the role and influence of feeding. Much of this vital research, however, remains somewhat hidden away in the scientific literature with relatively little being translated for the rest of us. This body of work now allows us to be critical in our assessment of these serious questions and of the various answers that are available.

This book has been written with the objective of increasing understanding of wild bird feeding and promoting best practice for the benefit of both the people and the birds involved. There are, of course, some people who are utterly opposed to this practice and have advocated for it to be stopped and banned. Some of this opposition is well reasoned and championed by knowledgeable and committed people with a genuine concern for the welfare of the birds. And some is otherwise.

In 2002 the *Wall Street Journal* became an unexpected participant in the discussions of bird feeding with the publication of a front-page article by staff journalist James P. Sterba.[47] Published on 27 December (while millions of Americans were participating in the Audubon Society's Christmas Bird Count), Sterba's title left little to the imagination:

> American Backyard Feeders May Do Harm to Wild Birds

With the subtitle:

> Feeding Wild Birds Lures Pests, Predators, Causing Illness and Distorting Populations

If you can, I urge you to track down this article, not because of the quality of the writing and certainly not as an exemplar of excellence in investigative journalism. Rather, I rate this as the highest-profile summary of the main arguments typically fielded against wild bird feeding. While there were plenty of impassioned responses to the article at the time (notably from John Fitzpatrick and André Dhondt of the Cornell University Laboratory of Ornithology, and blogger Laura Erickson),[48] considering the issues articulated in this article provides a suitable guide to whether our understanding has improved in the time since the article appeared. In effect, this entire book is an attempt to evaluate and respond to the issues—some legitimate, some just provocative—raised by Sterba's piece. The outcome will be a guide to how much we really know and how much we don't.

Where Are We Going with This?

We started this journey with the arrival of those birds at my table. As we will see, there are lots of reasons why we feed wild birds, but at the heart of this activity is the experience. It is something about the participating, the interaction, as well as the helping and learning. These complex motivations will be explored in Chapter 8 ("Reasons Why We Feed"), describing some of the most recent and illuminating studies to have been conducted in this field.

First, though, we must attempt to discern how and why people started to provide food for wild birds, and how these humble beginnings led to the global bird-food industry we have today (Chapter 2: "Crumbs to Corporations"). This examination of the history of feeding uncovered a major change in the practice of feeding, with the move from winter-only to year-round feeding in some places (Chapter 3: "The Big Change"). Having explored the scale and pattern of feeding, we ask whether all those feeders are affecting the distribution and behavior of birds (Chapter 4: "The Feeder Effect"). This is followed by a detailed account of how scientific studies of supplementary feeding are transforming what is known about the role of food in the life of birds (Chapter 5: "What Happens When We Feed?")

The material covered to this point should provide the background necessary to tackle the controversial issues raised above. The crucial issues of disease and nutrition are explored with some trepidation in the next section (Chapter 6: "Tainted Table?"). Beyond the backyard, the provision of supplementary foods has been successfully employed in conservation programs throughout the world, sometimes with unexpected results (Chapter 7: "Feeding for a Purpose").

But what about the people? Why do we feed birds anyway? Thankfully, a number of extremely important new studies have explored the many and complex reasons behind our participation in feeding. This is explored in Chapter 8 ("Reasons Why We Feed").

By the end, what have we learned? And does it really matter anyway? In the final section (Chapter 9: "Bird Feeding Matters Even More Now") I will argue that feeding is now a fundamentally significant component of the urban landscape where most of us live. This brings great opportunities as well as genuine risks.

2

Crumbs to Corporations

The Extraordinary History and Growth of Bird Feeding

Shrewsbury in Shropshire, England, is a rather unimposing place to have such a prominent role in our story. It was a dull, overcast, and drizzly day when I visited in early November. It seemed to be somewhat out of the way, up in the remoter parts of the northern Midlands, yet despite driving along a maze of minor roads in a quiet rural landscape, we passed a steady succession of huge trucks heading in the same direction. We (and the trucks) were on our way to the massive CJ Wildlife complex, the headquarters of one of the largest bird-food companies in Europe. My guide was colleague and collaborator Jim Reynolds from the University of Birmingham, one of the numerous authorities I was visiting to talk about bird feeding in Great Britain. Jim has connections across the world as well as locally and had generously suggested I accompany him to Shrewsbury, about an hour's drive from Birmingham. I hoped such a trip might be possible: the CJ factory ships vast amounts of birdseed and feeding hardware throughout Great Britain and increasingly to Europe. Seeing this hub of the birdseed industry would provide an important perspective. But first, I learned suddenly, we were to visit Mrs. Irene Cuthill for a "quick cuppa."

As we pulled up to Mrs. Cuthill's humble house on the outskirts of Shrewsbury, Jim said, "Right on time for Irene's famous scones. But this visit was much more than a tea break. I would like you to see something I think is central to understanding this whole bird-feeding phenomenon," he explained as we waited at the front door. Irene appeared wearing a gray cardigan and floral apron. Over the following hour, she proceeded to dismantle any expectations I might have had about an English widow who fed the birds.

Typical of many streets built in prewar Britain, Irene's front yard was minute although her backyard was more substantial, and even featured an ancient oak tree. "No idea how old it is," Irene mused, "but it's truly precious. When the children were here they lived in that tree. When they moved out, I thought, Let's see if I can help another 'family' in some way. That year we had a lot of snow and the birds just looked desperate. All I did was put up a feeding tray, but in no time I had a few regular visitors: tits and sparrows, and robins on the ground below. It was nice to see them come, but they didn't stay for long each visit. Something got me thinking about the garden from the bird's point of view. And that's when I realized just how bare it was. There was nowhere to hide or rest." This was a revelation that led to big changes in that modest backyard. She added shrubs, a pile of dead branches, and some small berry-bearing trees, all her own idea. "I learned much later that what I was doing was called 'wildlife gardening'; I just called it common sense." And with growing enthusiasm came a willingness to try some of the brand-new products that were then starting to appear in the shops: fancy see-through hanging feeders, ready-made peanut cakes, and a strange tiny black seed that the locals called "thistle."

She didn't think much about it at the time, but when a series of unexpected species started to turn up, she began to wonder whether her place might just be a bit special. At first it was a few Siskins, then more and more Goldfinches, and recently even Blackcaps. "I knew our yard was rather different from most around here; sometimes I think people thought us a bit queer, with all these plants and feeders. But when these strange birds appeared, I thought, Blow the neighbors! It seems that the birds like something about the place. I think the secret was the food *and* the shelter."

"I used to feed only in winter," Irene explained. "I made my own suet balls from the Sunday roast lard and hung them in the trees. I started with the first frosts, usually, and stopped with the first signs of spring. We all knew that the birds needed to be able to look after themselves; that they might get too used to the handouts. But I reckoned that the birds are smarter than that. In October they hardly touch the feeders, they're off in the woods feasting on berries." It was a gradual thing, but Irene soon realized that she was providing a little food for most of the year. "I was a bit worried about it at first, but the birds seem perfectly happy." Her hesitancy about feeding in the warmer months was finally put to rest when she was watching *Springwatch* on the BBC. "That funny chap, you know, was explained that feeding was completely fine at any time of the year. That was good enough for me!" Jim saw my blank expression. "Bill Oddie," he explained.

For a keen bird watcher eager to see the birds up close, the increasing availability of various products was also a key element in the move away from winter-only feeding. "In the beginning I only fed homemade things: fat balls, kitchen scraps, and the odd handful of crumbs. And now I can get all these wonderful seed mixes, made just for birds, in the local supermarket! Jim tells me that all this stuff is made right here in Shrewsbury! What a world we live in!"

Jim was right. A short if convoluted drive away, is the vast complex of CJ Wildlife (address: The Rea, Upton Magna). Known until recently as CJ Wildbird Foods Ltd., the company has broadened its view to include a comprehensive range of products generally associated with promoting nature in urban areas (hence their new advertising motto: Bring more wildlife to your garden). The spectacular growth and diversification of this company from its origins in 1987 as a routine "sunflowers" packager to one of Europe's largest wild-bird-food producers is a powerful example of the development of what is now a huge international industry. According to Chris Whittles, founder of CJs, the company he sold recently was making £19 million annually and, in addition to its main market in the UK, now operates in nine other European countries.[1]

The CJs (the letters are from Chris J. Whittles's initials) complex could be just about any major enterprise engaged in large-scale transportation of goods, possibly something associated with rural enterprise given its location on a working farm. A series of massive anonymous-looking warehouses encircle a much older brick building from which rumbling noises emerge continuously. Huge trucks are parked in various docks, unloading or receiving, before thundering off down the country road. It is not the kind of place you would feel comfortable wandering around without a guide. Thankfully, we were soon greeted by Martin George, a long-term employee of CJs and a recent graduate of the ornithology course Jim runs at Birmingham. Martin's intimate knowledge and insights concerning every aspect of the operation made the tour a fascinating exploration of the inner workings of the bird food machine.

I had assumed that a birdseed factory would be little more than a giant automated packaging machine. Not this place. From its earliest days, CJs made the decision to stand out from similar enterprises through a commitment to the highest quality in everything. Everyone knows what cheap, poor-quality birdseed looks (and often smells) like. Genuine quality starts with careful selection of suppliers and continuous assessment of the seed itself. As soon as the raw materials—the sunflower, corn, nyger, wheat, millet—arrive at the factory, all manner of checking, monitoring (of moisture levels, fungi, and many other factors), and cleansing techniques are undertaken: the rumbling sounds we could hear were the routine air blasting of all arriving seeds to remove dust and detritus. Deeper within the red-brick building we watched the massive mixing machines meticulously combining (in this case) sunflowers hearts and pieces, corn bits, and other ingredients in the same recipe developed by Chris Whittles in the early days. The resulting Hi-Energy No Mess seed mix, still the biggest seller in the ever-expanding array of specialist products, was being collected in 20-kilogram bags by the secure hands of two workers. A large proportion of the 250-strong workforce were locals, many with a farming background and an affinity for hard work.

In one of the other vast warehouses Martin showed us smaller automatic bagging machines deftly sealing seed in exquisitely designed 1-kilogram bags (in this case, it's Less Mess Sunflowers; they seem to have a knack for clear, no-nonsense product names). An immediately evident feature of all the products is the visual impact of the packaging; these packs feature high-resolution bird photographs and a see-through panel so the prospective purchaser can see what is within. The bags we observed happened to be labeled in Dutch, destined for the giant European distribution plant in the Netherlands; others were in German, Swedish, or Polish. In the next building we walked past seemingly endless storage shelves stacked to the lofty ceiling with every conceivable piece of feeding and wildlife gardening hardware—feeders, nest boxes, feeder stands, squirrel guards, even squirrel feeders—as well as the latest range of ultra-specialized products (my personal favorite: Organic Pate for Hedgehogs). Martin's enthusiasm was infectious but the sheer scale and diversity was, frankly, overwhelming.

And while CJs may have been pioneers in the UK, these days they are certainly not alone. Competitors include Haith's, Vine House Farm, Jacobi Jayne, Westland, Gardman, Henry Bell, and plenty more.[2] Aggressive expansion and competition have been characteristic of the 2000s and show little sign of slowing. Reliable data for the UK are surprisingly difficult to find, but one estimate is that the industry turned over £550 million in 2009 and continues to grow.[3]

Across the Atlantic, the industry is both fiercer and more diverse. North American bird feed consumers are regarded as being well informed and selective, readily purchasing a bewildering array of separate products: species-specific seed mixes, squirrel guards, nutritional supplements, drinkers, and an astounding variety of feeders. These items are available from numerous outlets including large chain stores, mass merchandisers, and specialist stores: pet, aquarium, gardening, outdoor recreation, and, increasingly, birding outlets. Some successful examples of the latter are Duncraft (founded in 1952), Wild Birds Unlimited (starting in 1981), and Wild Bird Centers (first opening in 1985 in Washington, DC and now with one hundred stores around the country) to name just a few.[4] And the industry is still growing. According to data collected by the US Fish and Wildlife Service, bird food sales were worth US$2.6 billion in 2001, US$3.35 billion in 2006, and US$4.07 billion in 2011; that's almost a doubling in a decade. We can add a further US$969 million to the 2011 figure to account for all the "nonseed peripherals."[5]

This is now a truly global industry. The United Nations estimates that the international production of the main seed types—sunflowers, nyger, and safflower—is growing at around 4% annually.[6] The biggest markets are based in Europe and North America, and the key production countries, such as India, Myanmar, China, and eastern Europe, have been exporting bulk seed in staggering volumes.[7] The amount of birdseed (which these days is overwhelmingly intended for wild rather than caged birds) sold during 2007 in the United States alone has imaginatively been pictured as filling over 22,000 railway hopper cars stretching for 252 miles (405 kilometers).[8]

How did we come so far in such a short time? This is a remarkable transformation from private practice to mass consumer behavior, from domestic leftovers to a multitude of finely tuned commercial products. The first mass-marketed seed and feeder products only appeared during the 1960s, and were slow to sell. Why would people spend their cash on some new seed pack when table scraps cost nothing? Why, in fact, did Mrs. Cuthill feel moved to start buying nyger seed and, fascinatingly, begin to feed her birds year round? It's an intriguing tale of ethics, habits, marketing, science, hard sell, desperate conservation, and the mysteries of consumer behavior.

The Origins of Bird Feeding

If you really want to discover something about the beginnings of bird feeding you need to go beyond a typical Internet search. While that is the obvious way to start, you quickly find yourself in the realm of endlessly repeated anecdotes and definitive pronouncements from all sorts of ardent but not necessarily sound correspondents. The frustration of reading the same stories and statements with slight but important variations eventually led me to seek out many of the original sources, thereby opening up unexpected side-stories and red herrings. Such is the tedium and joy of being an archaeologist-cum-detective, discovering what the original author actually wrote and uncovering entirely new components of the story. As this excavation proceeded, it became clear that there was a genuine history to be described here. "Fieldwork" was required, beyond Google and Wikipedia, into the strange landscapes of the restricted entry sections of big libraries, early editions of long-defunct newspapers, old naturalists' journals, and shared notes with colleagues from around the world. As has so often been the case, by far the bulk of what I was able to find comes from the same familiar places, namely the United Kingdom and the United States. Again, this will bias the story because I am also aware that bird feeding emerged with similar patterns in other countries, especially in Europe, but finding written historical material has proved elusive. The key exception is Germany, which has had an important influence on the development of bird feeding practices.

It is also important to acknowledge some of the key sources of information on which the narrative that follows is based. In the absence of a published account of the evolution of the industry in Britain, I am strongly indebted to Mr. Chris Whittles,[9] whose personal perspectives were invaluable. In welcome contrast, the story for North America is now available in unusual detail in two impressive compilations written by Paul Baicich, Margaret Barker, and Carrol Henderson: *Feeding Wild Birds: A Short History in America* (2010) and the much more detailed *Feeding Wild Birds in America* (2015).[10] These works were of immense value to this project, and I have borrowed heavily—but not uncritically—from them. It should also be stated that these sources are hardly disinterested reviewers: undeniably their views are those of industry insiders. The *Feeding Wild Birds* booklet, for instance, was commissioned by Wild Bird Centers of America, Inc., one of the vigorous specialist bird franchises in North America. For British and European materials I am also greatly indebted to David Clark from London (another graduate of the Birmingham ornithology course) for his forensic scrutiny of many early newspaper sources.

Some of the older stories unearthed are fascinating, and I am happy to retell the more colorful and informative ones. Many, however, are not really about the origins of bird feeding per se; they are often examples of something the writer found to be unusual or worthy of comment, a cultural practice or a description of a piece of feeding equipment. What we are more interested in here are the practices and ideas that led to the modern approach—organized, planned, costing something—to wild bird feeding around our homes. But that will become clearer later.

Having warned against definitive statements, let me make one of my own. It may be provocative but is essential for where we are heading.

> The spontaneous offering of food to wild birds is probably global.

Wherever fries ("chips") are thrown to the gulls at the beach or bread is offered to ducks at the pond, when we feed the pigeons in the town square or spread crumbs on the snow-covered windowsill, the impetus is the same. It is the unspoken, somewhat mysterious—almost innate—response to an animal seeking something to eat or caught in a potentially stressful situation. Not everyone responds, but perhaps most people do. This inclination, if not universal, is certainly cross-cultural. I have seen Bornean women leaving out leftover rice for Fireback Pheasants, Nepali Sherpas tossing crusts to Spotted Great Rosefinches high in the Himalayas, and Inuit hunters throwing seal fat to Snowy Owls. In the process of researching this book people from around the world told me stories of typical and spontaneous bird feeding: bread or grain or fruit tossed to West Peruvian Doves in Lima, Peru; Brehm's Tiger Parrots in the Central Highlands of Papua New Guinea; Wattled Honeyeaters in the Samoan Islands; Namaqua Doves in Khartoum, Sudan. While the species and locations change, the essential elements of these interactions don't. People everywhere tend to respond to the approach of a wild bird by sharing their food. This reaction also seems to be more likely if there is obvious need (such as during harsh weather) or the creature is small and cute or "special" (rare, unusual, albino), but not necessarily.

My point is that this form of spontaneous wild bird feeding is common, normal, and possibly instinctual. To search for its origins is simple folly. I would contend that it was a response present among our earliest human ancestors. It may even be regarded as an indication of authentic recent human evolution, along with conceptual language, symbolic art, and religious practice. Something significant occurred the day some distant ancestor offered sustenance to the sparrows hopping around the campsite rather than seeing them as hors d'oeuvres.

Believe it or not, this bit of pseudoanthropology actually helps clarify our quest for the origins of contemporary wild bird feeding. The stories and anecdotes may be fascinating and potentially edifying but they tell us little about where bird tables came from, why tube feeders are important, and why people went out and bought them in the first place. People everywhere opportunistically feed birds that they encounter, but not everyone does so regularly, using purpose-made or purchased items deliberately to help attract them closer. I would describe this style of feeding as:

> Systematic, intentional provisioning of foods for wild birds involving a level of planning and tangible cost, either in time or money or both.

This type of bird feeding is deliberate, planned and purposeful. As shorthand, I will call it "organized feeding" to distinguish it from the simpler "spontaneous feeding." It's the difference, for example, between feeding birds by discarding the crumbs from the table and the regular, planned provision of food prepared or purchased specifically for that purpose. This distinction became clear to me while investigating duck feeding in a suburban park with my colleague Renee Chapman.[11] While a lot of people and waterfowl were interacting, most of the items offered were discards from picnic lunches. Even though the feeders had probably anticipated feeding, almost none of the items they distributed had been purchased for this purpose (we asked): this was spontaneous feeding. In contrast, a construction worker routinely came to the pond to eat his lunch (home-prepared sandwiches) but he also distributed cooked rice prepared specifically for feeding ducks: this was organized feeding. Our quest, therefore, is to see if we can discover how both versions of wild bird feeding have evolved and developed.

A Tentative History of Wild Bird Feeding

Birds feature everywhere in ancient cultural records, especially in the religious texts from many traditions. This is a rich seam followed in Mark Cocker's *Birds and People*,[12] the outstanding compendium of the way birds have featured in cultures throughout the world. Birds are depicted as spirit guides and intermediaries, villains and tricksters, agents of evil and even as deities (with corvids—crows and ravens—being mentioned remarkably often). Ancient and medieval writers also employed birds as metaphors, exemplars, and similes ("You will rise up on wings like eagles," Isaiah 40:31). Real birds—as opposed to literary devices or religious motifs—appear less often, and where they do, they often feature as game to be hunted, potentially dangerous wildlife, or occupants of remote or desolate locations.

As far as I have been able to determine, the very earliest mention of the feeding of wild birds is found in Hindu writings of the Vedic era, at least 3500 years ago. These texts describe the daily requirement for orthodox Hindus to practice *bhutayajna*, one of the *panchamahayajnas*, the "five great sacrifices" designed to mitigate the accumulation of negative karma.[13] The *bhutayajna* stipulates the provision of food, traditionally rice cakes, for birds but also "dogs, insects, wandering outcasts, and beings of the invisible worlds." Given that this remains a standard practice of many contemporary Hindus, it surely is the longest running form of organized wild bird feeding.

No civilization can claim a stronger relationship between birds and its religious life, however, than that of the ancient Egyptians.[14] While a number of species feature in Egyptian writings and rituals, as divine representatives on earth or as metaphors for divine attributes, two species, the Sacred Ibis and the Peregrine Falcon, predominate in this spiritual landscape. The vast numbers of ibis (sacred to the god Thoth) mummies involved (Saqqâra alone holds 1.5 million; several sites were capable of processing 10,000 birds annually) have been well documented, but less well known are the millions of falcons (representatives of Horus) that were employed in a similar fashion.[15]

An obvious logistical question arises: How did the Egyptians acquire the birds needed in such numbers? We know through ancient administrative texts that both species were raised specifically in captivity for such purposes as well as being harvested in huge numbers from the wild. To enhance the steady demand for falcons, a stipend was provided by the royal household to the priests to be used for the maintenance of fields dedicated to provisioning falcons with food; a statue commemorating a man named Djedhor describes how he "prepared the food of the falcons living in the land." Similarly, fields were set aside for exclusive use by ibis and were overseen by priestly wardens. Dating from about 700 BCE, this must surely be the earliest form of mass, well-organized, planned bird feeding. This was intentional provisioning for the living birds; when they were dedicated (which involved capture, ritual killing, and mummification), food was also provided for their journey accompanying the deceased to the afterlife: recent X-ray examinations of ibis mummies have discovered special foods inserted into their bills during preparation.[16]

Within the Judeo-Christian tradition, the earliest writings possibly associated with birds and feeding are thought to be certain passages from the book of Leviticus (written around 1440 BCE). Among the various laws proclaimed is an admonition for some of the harvest—the grain growing at the edges of the field and the fallen gleanings—to be left in place "for the poor and the foreigner among you" (Leviticus 23:9). To this list of unfortunates some scholars have added birds, although this has been contested. A much more characteristic theme is found in the

New Testament, in the gospels Luke and Matthew (ca. 80s or 90s CE), of God's benevolence and care as exemplified by his provision of food for the birds (for example: "Consider the ravens; they neither sow nor reap, they have neither storehouse nor barn, and yet God feeds them" (Luke 12:24). This is a powerful image: God as bird feeder, who cares even for the lowly sparrow (Matthew 10:29). Elsewhere in the New Testament, birds are portrayed as agents of wasted opportunity in Jesus's parable of the sower (Matthew 13:4), consuming the metaphorical "seed as the Good News" carelessly cast onto the path rather than onto the tilled soil. These would have been effective metaphors for Jesus's rural audience, who would have been familiar with the local birds ready to scavenge any seed they could. For our current purposes, this reminds us of those instances when bird feeding occurs against our wishes: the unwelcome species at the feeder; those aggressive waterbirds that invade the picnic; the scavengers of human food wastes.

A final biblical example of a feeding interaction disturbingly reverses the expected arrangements: God directed the prophet Elijah to await further instructions from a cave in the dry and remote Kerith Ravine, east of the Jordan River. How can he possibly survive? By wild bird feeding with a difference. "Ravens brought him bread and meat in the morning and bread and meat in the evening" (1 Kings 17:6; ca. 550 BCE). How's that for deliberate, systematic, and regular provisioning of species-appropriate sustenance?

In reality, however, the historical record—at least the component available in English—is strangely silent (ignoring the Egyptians for the moment) about what we would accept as just about any form of bird feeding from the first century CE until somewhere in the eighteenth. Maybe other things were happening—the Dark Ages, the Crusades, the Reformation—but writers, philosophers, and journalists seem to have missed the feeding undoubtedly occurring in their very own streets and villages. In all seriousness, that is the most likely explanation: it was so familiar and commonplace—so ordinary—as to be unworthy of comment.

There are a few worthy exceptions to this dearth of historical detail, although their veracity may be questionable, both involving Roman Catholic saints. The first features the somewhat opaque Scottish figure Saint Serf (or Serbán) (ca. 500–583) of Fife.[17] Among numerous highly improbable adventures (including seven years as pope in Rome) and the usual series of miracles, it was his apparent "taming of a wild robin by the act of hand feeding" that has often warranted mention. Although not directly related to feeding, Saint Cuthbert of Northumberland (634–687) also deserves attention here in the context of a very early concern for bird conservation. Arguably the most famous saint of Anglo-Saxon England, Cuthbert is today recognized for enacting the world's first bird-protection laws. During a spiritual retreat on the nearby Farne Islands, Saint Cuthbert used his authority as bishop of Lindisfarne to declare legal protection for the eider ducks and other seabirds that were being harvested unsustainably by fishermen.[18] These laws—literally centuries ahead of their time—remain in place today.

Saint Francis of Assisi[19] (ca. 1181–1226) is venerated for his revolutionary ideas on many topics, but of particular relevance here is his conception of the relationship between humanity and nature. Francis regarded the natural world as "the mirror of God," and therefore all animals were fellow creatures to be treated with appropriate respect. He famously preached to flocks of birds gathered expectantly beside the road—although there is no mention of him actually feeding them. He does, however, convince some irate villagers to feed a starving wolf instead of killing it. The legend says they did so.

The long slow centuries without much reference to bird feeding come to a whimsical end with the advent of the era of broad circulation newspapers, especially in England. For example, on one apparently slow news day in 1787, the *Northampton Mercury* felt it "worthy of Remark," that a "Pair of wild Sparrows have built a Nest and hatched their Eggs in the kitchen," and that the "Mistress of the House often feeds the young Ones."[20] Furthermore, a predilection to bird feeding may be an indicator of moral character according to a character reference tended to a Scottish court. The accused murderer, according to an acquaintance, was a "kind and mild man of a sensitive nature. He used to carry crumbs of bread for the purpose of feeding birds."[21] We do not know whether this swayed the jury.

Scrounging for Crumbs

Almost a century later in the United States, Henry David Thoreau (1817–1862) is detailing his observations of nature and philosophy during his experience of influential solitude spent in the woods at Walden Pond in

Massachusetts. His masterpiece, *Walden; or, Life in the Woods,*[22] covers a broad canvas but also describes his encounters with and impressions of a wide variety of animals. Published in 1854 but covering the years 1845–1846, the date of 1845 is widely cited as indicating that bird feeding was already under way in America.[23] The frequency with which this claim has been made provoked me to reread a book that moved and inspired me as a young ecologist many years previously. Immersing myself once again in the beautiful prose with more mature (and possibly more cynical) eyes was bracing and refreshing but also heartbreaking as he describes a world now lost forever. But also hopeful because his experiences are equally concerned with our personal attitude as much as place: anyone, anywhere can appreciate whatever element of nature is available, from the inner-city apartment balcony to the solitary hut in the woods. For Thoreau, this required a trained eye and a patient soul.

Walden is a formidably honest exploration of a personal relationship with nature on its own terms. Thoreau quotes an ancient Hindu proverb: "An abode without birds is like a meat without seasoning" and goes on to say, "Such was not my abode, for I found myself suddenly neighbors to the birds; not by having imprisoned one, but by having caged myself near them." His encounters with a vast array of species—many now rare or gone—fill the book with wonderfully moving stories. But what of feeding? Where does the significance of the claim of 1845 lay? There are descriptions of a French-Canadian visitor sharing his fire-cooked potato with a wild chickadee and of Thoreau himself hand-feeding cheese to red squirrels and even a rat. Finally, I found the following observations:

> In the course of the winter I threw out half a bushel of ears of sweet corn, which had not got ripe, on to the snow-crust by my door, and was amused by watching the motions of the animals [squirrels, jays, and chickadees] which were baited by it.
>
> A little flock of these titmice came daily to pick a dinner from my woodpile, or the crumbs at my door. (183–184)

That, as far as I can ascertain, is it; statements of such magnitude that the year 1845 is nominated as the first mention of wild bird feeding in the United States. In the absence of other published accounts, this may be technically accurate. For our present purposes, however, it provides little beyond the rather obvious proposition that New Englanders, like everyone else, were also tossing scraps to visiting birds.

Whatever the significance of that year, things were indeed beginning to change during the mid- to late 1800s, on both sides of the Atlantic. The first tangible signs of organized, systematic, even commercial wild bird feeding began to emerge during this period.[24] Perhaps surprisingly for the times, numerous US women were early advocates of feeding wild birds, in large part as an alternative to the accelerating hunting, trapping, and deliberate killing of birds. Titles such as *Birdcraft* (1985) by Mable Osgood Wright and *Birds of Village and Field* (1898) by Florence Merriam Bailey provided detailed observations as well as practical instructions on how and what to feed birds. The latter includes a description of a Mrs. Davenport who made her own "bird bread" of one-third wheat and two-thirds Indian meal (a mixture that did not freeze) and presented it on a special feeding station.[25] These are undeniable indications of organized bird feeding. It was certainly not common yet but was definitely on its way.

The publication mentioned earlier—*Feeding Wild Birds*—makes a strong case for the impetus of such early promotion of feeding in North America as being related directly to the fledgling bird-protection movement, itself a reaction to the colossal scale of bird slaughter and waste associated with the fashion plumes industry.[26] Peaking around the turn of the twentieth century, untold numbers (5 million in 1885 alone) of birds, particularly egrets, were collected for stylish ladies' bonnets, although entire woodpeckers and dead sparrows were also commonly embroidered on many a high-society blouse. Organized opposition to this destruction led directly to the formation of organizations such as the Audubon Society and the eventual enactment of bird-protection laws and statutes. Feeding rather than killing birds was advocated from the first edition of Audubon's new *Bird-Lore* magazine in 1899, with Isabel Eaton encouraging teachers in particular to nurture their pupils' interests in nature study by "coaxing [the birds] onto the window shelf with a free lunch."[27]

The First Feeding Devices

A definite contender for physical evidence of the start of modern bird feeding in the United Kingdom would have to be the *Ornithotrophe*, the world's first wild bird feeding device (and operated only eighteen miles away from Shrewsbury). Invented by John Freeman Dovaston (1782–1854) from

Shropshire in 1825,[28] this construction (he described it as a "bird feeding trencher") was a modified wooden cattle trough fitted with rows of parallel perches and filled with a variety of domestic and farmyard food scraps. In his meticulous notes, Dovaston recorded twenty-three species of birds using his device over the winter. Unfortunately, it appears that the *Ornithotrophe* did not catch on, with subsequent references to the device disappearing from all records.[29]

As the Industrial Revolution rolled on in Britain and Europe, significant sections of society in North America and Europe were becoming relatively wealthy and increasingly urbanized. The importance of the private garden as an indicator of affluence and as place of leisure and refuge was a feature of this era. In David Callahan's fascinating *History of Birdwatching in 100 Objects*[30] (a replica of Neil MacGregor's slightly more ambitious *A History of the World in 100 Objects*) there are several items of direct relevance to our present discussion, the first being the advent of the "wildlife garden." Callahan identifies the start of this movement as circa 1835 in England and its flourishing during the long Victorian era that followed. As well as providing an aesthetically tasteful setting to escape from the industrial landscapes that characterized the era, a growing number of private landholders began to perceive of their lands as more than pleasure gardens. Initially associated with larger estates that had the financial capacity for extensive plantings and maintenance, the defining feature of these wildlife gardens was the provision of habitat. This could be the revegetation of denuded waterways or simply the addition of bird-friendly shrubs in a typical smaller garden. At this stage, commercial feeders would have been unknown but birdbaths, either the fashionable stoneware pedicel model or the improvised washbasin on a stand version, became a standard feature.

Within twenty years, however, the bird table had made its first appearance. The prototype of the familiar garden "little house" design appears around 1850 according to Callahan,[31] providing the basic horizontal platform for bird food, sometimes with some sort of a roof for protection from the weather. At this time the construction was still very much do-it-yourself; the widely available commercial models were almost a century away, although simple plans for various designs were in general circulation. The food presented would also have been homemade or homegrown. Grain, if any, would have been originally intended for the domestic fowl but more likely was the widely cited "scraps": vegetable peelings, table leftovers, and spoiled or stale food. These were the days when, for the great majority, little would have been discarded as waste; feeding birds with kitchen scraps would have appeared sensible and probably virtuous. Nonetheless, the pattern of provisioning was fairly sporadic, opportunistic, and—for the birds—unpredictable.

Climatic Cataclysms

Reporting on the impact of extreme weather events on birds had long been a feature of the British press. In February 1776, the *Ipswich Journal* noted during one prolonged winter storm:

> The quantity of fowls of different kinds caught up in this season is incredible, owing to the hardship they are exposed to for want of food; finding their usual supplies locked up from them, the hardy bird of prey is taken by hand, while the sweet linnet and sparrow are cramped with cold and die. The poor seagulls, in great flights, are displaying their token of necessity, by approaching the houses, and scrambling for the crumbs at the door.[32]

These "crumbs," token offerings of assistance in a time of need, become ubiquitous in accounts from this period, as was feeding as an expected form of duty. An 1855 letter to the editor of the *Kendal Mercury* suggests the practice as common and compassionate:

> Sir,—The birds mentioned in my last letter were those which seek shelter, during severe weather, around the homesteads of man. Their presence in his windows is a sure sign of the failure of food; and generally some member of the house, pitying their wretched appearance, has compassion enough to offer them a more bountiful supply than they can gather from the mere crumbs about the doors. We should, indeed, be sorry to see this ancient season of feeding the robin,
>
> > "Thrive happy creature! In all lands
> > Nurtured by hospitable hands,"
>
> . . . from time immemorial, drop into disuse. To a child this duty of casting a few crumbs of bread on the windowsill, we may look upon as an early lesson in morality. Thus may our child's sympathy for a fellow-being in distress, be early excited, and tutored by the agreeable occupation of feeding birds in winter.[33]

There are also fleeting references to a more community-based form of feeding. The *Edinburgh Evening News* of February 1875 mentions that a "society of feeding birds in the winter" has established twenty-two feeding stations in the town, providing three meals per day.[34] The expense incurred, it is explained, will be "repaid a hundredfold by the destruction of injurious insects." Such collective action would certainly be a significant development for this narrative, but unfortunately no further details have come to light as yet.

It is during this period that one of the seemingly defining events in our story—"an apparently pivotal moment" according to David Callahan[35]—occurs in Europe. The 1890–1891 winter was early, long, and brutal, affecting all of Europe and northern Africa. In terms of scale of impact and the temperatures recorded, that winter is regarded as the most severe in the nineteenth century.[36] In the UK, the worst conditions were experienced during the Great Blizzard of 9–12 March 1891. Extensive snowdrifts and powerful storms swept repeatedly through the south and west of England, sinking fourteen ships and resulting in the loss of 220 lives. While conditions were clearly life-threatening for many humans, somehow, in the trauma that followed, the plight of wild birds was brought to the public's attention by the nation's many newspapers. Correspondents from the countryside and the cities, appalled by the toll ("Under hedges and bushes their dead bodies, consisting of not more than feathers, skin and bones, are frequently to be seen"), described vividly the situation as they perceived it, and implored people everywhere to "reach out and help the birds."[37] Recipes, suggestions, and advice were published widely and appear to have been heeded by so many that we may regard this event as marking the start of winter feeding at a national scale, in England at least.[38] Spontaneous, opportunistic feeding continued as always; this new mode was, however, deliberate, regular, and, to an important extent, planned.

The drama of the times can be glimpsed through an exchange of letters to the *London Daily News* written in the midst of the 1890–1891 storms (30 December 1890):

> Sir,—I am much obliged for your kindness in reminding your readers of their duties to our friends, the birds. Their suffering is far greater than any can realise who are not frequenters of our woods and fields. It would be great service if some of your better-informed readers were to supply us a little more

> studiously with information as to the most suitable food, and the best mode of supplying it.[39]
>
> T. C. Williams

This heartfelt request is granted from an unexpected scribe:[40]

> Sir,—In the name of my brothers and sisters of every "wing," I thank you for the kind manner in which you have provided for us in this time of bitter cold and hunger. We can only promise to show our gratitude to the kind friends who have given us food that when next summer comes again, we will sing our very best songs and gobble up every worm and grub that we can possibly find.
>
> I remark that there is a division of opinion on what is the best form in which to feed us. Now a very dear friend of ours makes up (with her own hand, I believe) a delightful compound with which she feeds us three times a day. Made as this recipe: Half a stale loaf or any odd pieces of bread reduced to a pulp with warm water. Add more water and stir in oatmeal or barley meal or both, and then a few handfuls of hempseed. This mixture made into a thick stiff paste which we can all sup with our bills, and the small-fry—those perky tits, chaffinches, sparrows, etc., which abound everywhere, are equally delighted with the crumbs. That ill-bred, vulgar family of birds, the starlings, are most troublesome over this mixture, and the modest, gentle thrushes and blackbirds often have great difficulty in procuring any, but in the end all get a "pick" and very grateful we are.
>
> I am, your grateful but starving,
>
> Johnnie Thrush

From such evocative prose was seemingly born a profound and, to an important extent, permanent change in feeding practice.

It is worth dwelling on this event for a moment. As the historian David Elliston Allen explains,[41] prior to this blizzard such concerns would have been generally regarded as thoroughly unacceptable. "Worse, it infringed the Victorian domestic code. In that Golden Age of Home Economy waste was an anathema. Stale bread was either saved for puddings or converted into breadcrumbs. The place for other scraps was in the stock pot for soup." Yet, the "long frost of 1890–91 finally seems to have broken down resistance." From this point, at least in the towns, the habit caught on. Only a few years later, Allen reports [during a prolonged cold period] "hundreds of working men and boys would take advantage of the free hour at dinner time to visit the bridges and embankments of the Thames to give scraps left from their meals to the birds."[42] One wonders what Her Majesty thought of such behavior.

The Baron

The next milestone on our journey would better be marked by a roadside monument, grandiose and, although weathered and lichen encrusted, yet still conveying the triumph of reason and science. We can identify the year as 1908, but the period to which it refers covers the last decades of the nineteenth century and the early twentieth. This date refers to the appearance of the English translation of a treatise published in German the year before, eventually to be issued in six languages and which ran to twelve editions. This book, *How to Attract and Protect Wild Birds: A Full Description of Successful Methods,*[43] provides the most compelling evidence of the advent of a modern approach to bird feeding, as well as exemplifying the growing international interest in nature conservation and a more direct action by governments and citizens. The place of this work in the history of bird feeding definitely justifies our discussing it in a little more detail.

Although written by Martin Hiesemann, *How to Attract* was entirely focused on the extraordinary efforts of Baron Hans Freiherr von Berlepsch (1857–1933), a German nobleman who used his ancestral estate of Seebach in Thuringia, central Germany, to conduct experimental ornithology on a spectacular scale. Having a passion for both bird conservation and scientific research methods, Baron von Berlepsch spent over thirty years developing and, as he would say, "perfecting" his practical solutions to the perceived decline of birds. He published his findings in a series of scientific papers in national journals and synthesized these along with practical instructions in a book that appeared in 1899;[44] much of the material collated by Hiesemann is derived from these works. Von Berlepsch's approach was ruthlessly rational; a colleague, Dr. Kartent commented: "He owes his success to the fact that he carries his experiments from a purely scientific point of view, without sentimentality or exaggeration."[45] Nonetheless, he was no heartless utilitarian either. "We do not protect birds solely because they are useful, but chiefly from ethical and aesthetical reason, as birds give beauty and animation to Nature." His motivation was to devise practical and proven means that could be employed by the public to sustain beautiful and useful species of bird; it did help having a huge estate, well-trained staff, and presumably an inherited fortune at his disposal.

The Seebach estate covered 500 acres, 80% of which was relatively undisturbed woodland with the rest being ornamental parkland and a plantation of poplar and willow. Seemingly all of these lands were utilized in the baron's studies. The most important element of his experimentation was the installation of nest boxes throughout the estate. Today, nest boxes are a standard technique for research into the breeding of many species of birds, sometimes using large arrays of boxes. In Chapter 5, for example, I describe a study of tits that involved a staggering 800 nest boxes. Yet over a century earlier, von Berlepsch had erected a total of 2300, most attached to trees in the woods. Critically, these boxes were most definitely not your mail order "Blue Bird Standard" type box. Guided by the baron's own "design with Nature" principles, they were all-natural cuts of wild timber with internal cavities precisely replicating those of actual woodpecker nests. The title of the book was entirely accurate: it was full of directions, dimensions, and measurements showing the well-trained and equipped naturalist exactly *How To*, as well as helpful diagrams of "Right and Wrong Installations" and examples of "Worthless Imitation."

The eventual outcome of these decades of labor and development of von Berlepsch's "rational system for the protection of birds" came down to three fundamental activities: the creation of opportunities for breeding; "suppression of the enemies of birds"; and winter feeding. The first of these was addressed through the use of nest boxes and the wide-scale manipulation of vegetation to provide shelter for nests. The second—the deliberate and sustained eradication of what were mostly natural predators—will strike the contemporary reader as being seriously non-ecological in outlook, but we must be careful not to assess these attitudes from current perspectives. When the view of the day meant that "progress gives [Man] the right, or rather obliges him to restore the balance of Nature," and one's chief concern was the protection of songbirds, it made sense to offer rewards for the "capture and death" of all "mammals, sparrows, goshawks, sparrowhawks, jays, magpies, crows and shrikes." Common Blackbirds were included in this list "should they increase at the expense of the nightingales."

It is, of course, the baron's approach to feeding that will be of the most interest here. As expected, he had some strong opinions. He was typically unsympathetic to those he regarded as irrational and sentimental: "Kind-hearted people have always taken pity on our feathered guests. But often the results have been out of proportion to the means employed. [They] feel satisfied and proudly conscious of having done a good deed; they do not pay attention to the fact that they have in no way relieved the birds." And when should feeding occur? "Birds *only* need feeding during and after certain changes in the weather, especially during blizzards and intense frost." For von Berlepsch, feeding was primarily a rescue mission, a temporary but vital form of assistance when small birds were at their most vulnerable. For the baron, this meant winter only.

That's the when; what about the how? It is in the design and operation of his four feeding "appliances" that we see the baron at his most innovative and instructive—and pedantic. His "Food-Bell" and "Food-House" are designs instantly familiar to the modern feeder, being obvious precursors of hopper and roofed platform feeders, respectively. What is startlingly unexpected is their sophisticated, industrial, somewhat overdesigned appearance. These are clearly well-advanced models, not early prototypes, strongly suggesting a period of preceding iterative development. To me, progressively looking for new "fossils" to add to the bird-feeding evolutionary tree, discovering the baron's "Food-Bell" was akin to unearthing a cell phone in a Neanderthal excavation. It's just too sudden, a quantum jump in complexity. And, as in the story of human evolution, it probably means that we have simply missed the numerous earlier stages in the developmental process. I probably just haven't been looking in the right sites (such as, for example, the magazines of naturalists' societies of the time).

Both the bell and house were meant for dispensing seed, with hempseed at that time the favorite, although the platforms of the latter could also provide fat, suet, and scraps of meat. Indeed, it was in the provisioning of a high-protein, high-fat supplementary diet that Baron von Berlespsch was at his most enterprising. His careful observations of the small birds of the German wood indicated that insect foods were vital, and his "Food-Tree" was an attempt to imitate a coniferous tree closely covered in insect eggs and larvae. This "appliance" was literally a small fir or spruce over which was ladled liberal amounts of a boiled mixture of the following: white bread (dried and ground), meat, hempseed, "maw" (I have been unable to confirm just what this is), poppy flour, white millet, oats, dried elderberries, sunflower seeds, and ant eggs, all combined in a large vat of beef or mutton fat. The instructions include copious advice on the best methods of heating, the need to employ two people for dispensing, and how to avoid scalding the tree while administering the hot mixture. Despite the "rank odors" emanating from this concoction, the educational benefits to pupils witnessing the eager feeding of many species meant, berated the baron, that all schools should install a Food-Tree on their grounds. Furthermore, householders without suitable trees were told they could simply add the same mixture to a length of wood with several bored holes: thus, the "Food-Stick."

Should the procurement of the necessary ingredients prove difficult, or the production of the concoction too onerous, European followers of von Berlepsch's methods could simply purchase the mixture in solid form from the firm of Hermann Scheid in Büren, Westphalia, under the commercial name of "Food-Stones." Indeed, most of the specialized nest boxes and feeding appliances mentioned in the book were available from several named manufacturers in Germany and Austria-Hungary. Indeed, Martin Hiesemann gently chides von Berlepsch for his reluctance to pursue patents for some of his inventions. In other words, by the early 1900s, a genuine and successful commercial industry based on wild bird feeding and the practical actions associated with bird protection were well entrenched in Germany at least. Numerous designs had been tested and production of several devices was under way. Furthermore, "Food-Stones" appear to be the first product specifically designed as wild bird food—as opposed to various seeds produced for other reasons—and available for sale. This had clearly occurred well before the appearance of *How to Attract and Protect Wild Birds* in 1908. It would appear that this date is indicative of the time that the information first became widely available to the English-speaking world. The significance of this publication to this version of history cannot be underestimated.

Demand for Advice

The arrival of this German book in the Anglosphere is significant as evidence of the growing interest internationally in natural history and nature preservation generally, and of practical applications such as bird feeding. People—at least those better educated and well-off—sought advice and direction. Above all, the main purpose of the work of von Berlepsch (and his effective scribe, Martin Hiesemann) was to promulgate a sensible guide to practical bird protection that was accessible to all. *How to Attract* continually reiterates the importance of hands-on, DIY solutions, providing all the plans, tips, and handyman guidance needed. The book's suggestions were readily adopted and adapted on both sides of the Atlantic.[46] We can state with some certainty that by the early decades of the twentieth century, the practice of wild bird feeding was becoming much more systematic and informed, although commercialization was still some way off.

In the United States, demand for practical natural-history information from the public led to the production of a growing body of private and institutional publications on topics such as game management, waterway restoration, and raptor suppression. Of direct relevance is *How to Attract Birds in Northeastern United States,*[47] Farmer's Bulletin no. 621 published by the US Department of Agriculture in 1914. The author, W. L. McAtee, is notable for advocating feeding platforms designed to be weatherproof and waste-proof with roofs and side ledges, and was one of the earliest to suggest utilizing hopper feeders such as are widely used for domestic fowl and game birds to the present day. Similarly, Edward Howe Forbush's widely cited 1918 work, *Food, Feeding and Drinking Appliances and Nesting Material to Attract Birds,*[48] provides clear evidence of a growing interest in enhancing the effectiveness of a practice that was growing in popularity. Forbush (1858–1929) was a prolific and well-respected New England government scientist who was appointed state ornithologist in 1908, with the chief purview of investigating the habits of birds useful to mankind. Produced as a circular from the Massachusetts Department of Agriculture, his publication draws directly from von Berlepsch (including reproducing precisely the famous Food-Tree recipe) but had been written mainly with the New England region in mind. It is also much more expansive and instructive than the German book, describing a dozen "appliances" known to be effective as seed dispensers, feeding platforms, birdbaths, and squirrel and sparrow retardants. The emphasis is on self-constructed devices, however; although some are apparently commercial items (e.g., the feed hopper and the Weathervane Food House), Forbush infers that homemade is the norm. Similarly, when discussing foods to be deployed in feeders, only "mixed birdseed," presumably sold for small cage birds, is mentioned. Rather, grains, seeds, nuts, and fruit that can be readily collected or even grown locally are recommended as practical for most people. Food intended for wild birds at this time was still sourced primarily from the domestic realm: cereals and stock foods, squash and pumpkin seeds, chicken eggs and eggshells, as well as by-products of the meat intended for human consumption—fat, rind, suet, lard, and marrow. Table scraps, of course, are mentioned, although white bread is regarded as a "poor food" (apparently the first time this issue is referred to). On the other hand, milk products such as cottage cheese and curds, are regarded as "excellent."

Forbush concludes his work with a dossier of foods preferred by almost fifty species, indicating the collation of a considerable amount of personal experimentation and a wide network of correspondents. He is also an early (perhaps the earliest on record) advocate for year-round feeding, suggesting that any site successfully attracting birds should be supplied continuously so "that [the birds] may find it at all times in case of sudden scarcity."[49]

Clearly, interest was growing, the demand for advice and assistance building quietly. Partly in reaction to the despoiling of the countryside by unrestrained industrialization, as well as a renewed appreciation of scenic landscapes, many people were beginning to express what we would probably now call "environmental consciousness," supporting the nascent national parks movement,[50] and calling for the preservation of birds, at least those regarded as pleasant and useful. The decades around the turn of the twentieth century witnessed unprecedented levels of political activity by well-organized groups of citizens coming together to promote and agitate for action. This was the context behind the founding of some of the largest and most successful environmental and conservation organizations still active today: the Royal Society for the Protection of Birds (1889), Sierra Club (1892), and the National Audubon Society (1905). These collectives resulted in declaration of the world's first national parks (Yellowstone in 1872 and Royal in Australia soon after in 1879) and eventually the first of many bird reserves in the UK.[51] Support for these activities was widespread, suggesting increasing attention to environmental issues, both local and national. It is likely that these were the people starting to feed birds, regularly and intentionally, at home. This was the embryonic market for wild bird feeding products.

The Products

We know from the Germans that some commercial items—nest boxes, seed dispensers, and even specialized bird food products (e.g., "Food-Stones")—were available for sale in some places in central Europe by the 1890s, although how many were sold is difficult to ascertain. By 1910, the magazine *Punch*[52] described wild bird feeding as a national pastime in Britain, and included several advertisements for mail-order feeders—though not yet seed. In 1913, possibly the first advertisement for feeding devices in North America appears in *Bird-Lore*[53] (November–December issue), highlighting the Dobson Automatic Sheltered Feeding Table. In what was certainly cutting-edge design for the time, this feeder—according to the ad—moved in the breeze like a weather vane, ensuring that the birds within were always sheltered. In an early combination of commercialization, conservation, and emotional marketing ("Do You Love Birds? Help Save Them")—something that will become increasingly familiar—this device was designed by, and available directly from, Joseph Dobson the director of the Illinois Audubon Society. It would be fascinating to know how many were sold.

From this point on, a steady range of feeding-related apparatus appear for sale in appropriate publications.[54] Nonetheless, it appears that the growing numbers of people engaged in feeding birds were doing so without too many purchased items. But bird feeding, still overwhelmingly a winter activity, was moving beyond the farm and backyard. During the 1920s, community bird-feeding programs began to emerge, typically run by schools, Boy Scouts groups, and local Audubon chapters. These programs organized groups of participants to stock feeding stations established in public parks during the winter, usually using suet. For example, in Meriden, Connecticut, in 1921, it was reported that two dozen people "by precept and example" had established winter feeding stations "to cover every section of town."[55]

In 1935, in the United States, the Northeast experienced a series of extreme winter storms, leading to a media-led campaign promoting the feeding of birds in a fashion reminiscent of the Great Blizzard of 1890–1891 in England. The pioneering bird guide author Roger Tory Peterson recalled how announcers on radio stations throughout the region implored people to feed the birds while ice covered the ground.[56] "For days scarcely a program on the air did not include an announcement about this. Everybody fed birds, from the fire escapes of New York City to isolated snowed-in farms in the back country." Peterson's revolutionary *Field Guide to the Birds*[57] had appeared only the year before (in 1934), arguably the single most important event in the history of bird watching. For the first time, ordinary people had a truly portable, easy to use, and comprehensive means of identifying the birds they encountered. Unquestionably, attracting birds to a feeder where they could be clearly viewed increased interest in feeding, which further facilitated observation and identification: a classic positive feedback mechanism. Peterson was acutely aware of this link, and from his earliest writings, actively promoted birding through feeding.

It was during the 1950s that the scene changed dramatically. In the United States, numerous seed companies, providing largely grain in bulk to the poultry and cage bird industries, slowly began introducing products packaged specifically for wild birds. The Kellogg Seed Company, for instance, which had been selling birdseed since 1918, began selling a seed mix based on choice experiments conducted by the National Audubon Society.[58] Labeled "Audubon Society Mixture," the launch of this product in 1953 was just the first of an endless series of collaborations between producers and environmental organizations. From this moment on, the line between science, conservation, marketing, and recreation becomes impossibly blurred.

A couple of entrepreneurial seed companies stand out during this crucial period of the 1950s. Knauf and Tesch (eventually to become Kaytee Co.) and Wagner Brothers Feed Corp., although competitors, collaborated effectively in developing the first mass-produced "birdseed" to be sold nationally through grocery stores. Blending ideally with postwar prosperity and modern expectations of convenience and availability, the transparent polypropylene 5- and 10-pound bags transformed the practice of feeding birds in the backyards of North America. Knauf and Tesch's first bags of wild bird food appeared in 1955, and Wagner's a year later. The latter also had another twist: a brand radically called "Four Seasons."[59] This product was effectively the first real shot in the "winter versus year-round" altercation. Similarly, across the Atlantic, a significant step in the enhancement of feeding occurred in 1958 with the arrival of the small, neat boxes of Swoop Wild Bird Food[60] (for 25 pence). A marketing triumph of the time, these convenient containers could be purchased in pet stores and department stores throughout the country, avoiding the usual untidiness of buying your seed in bulk. The era of convenience packaging, with its attendant marketing opportunities, had also arrived in Britain.

But seed needs a feeder, and along with the expansion of offerings, the variety of US devices available for sale also increased. Simple homemade platforms began to be replaced by imaginative and attractive feeder designs, ideal as accessories for the modern garden or backyard in the sprawling suburban landscape. Duncraft, for example, founded in 1952, produced a windowsill feeder, first of Masonite, then in 1958, the "Flight Deck," made out of plastic.[61]

Not withstanding von Berlepsch's futuristic Food-Bell appliance, it is fair to say that until the late 1960s, most wild bird foods were being presented on flat surfaces such as platform feeders, with or without roofs, or simply spread on the ground. Despite the obvious issues of spoiled food, hygiene, squirrel depredations, and the continuous requirement of replenishment, most people would simply have accepted these as avoidable and inevitable. Enter some humorous Yankees and their contender for most significant exhibit in this history of bird feeding.

Believe it or not, the now-iconic company Droll Yankees, started life in 1960 as a Rhode Island record company dedicated, in the words of founders Peter Kilham and Alan Bemis, to "preserving the off-color humor and tales of Old New England."[62] Whatever the success of this venture, Kilham, a keen amateur birder who was also a trained engineer, began experimenting with alternative approaches to the delivery of seed. His triumph, the original tubular bird feeder (technically, model A-6F, but often regarded as *the* hanging tube feeder) was introduced to the public in 1969. Instantly, or so it seemed, the perennial issues of mess, waste, and efficiency were largely solved. This new vertical perspective also suited the emerging opportunities associated with the arrival of new seed varieties and the move to targeted feeding. The new tube design allowed far more control over access to the seeds through careful sizing and positioning of the feeding holes. The other feeder designs did not disappear, of course, but the tube feeder was one of those "instant hits," selling in huge numbers, and was a catalyst for experimentation in feeder design that has continued ever since.

The Seeds

The gravity-powered tube design was ideally suited to the relatively large and smooth gray-striped sunflower seeds that had become the predominant commercial bird food at the time. Prior to the 1930s, when sunflowers were first tentatively offered for sale as wild bird food, most of the seeds used were those readily available for agriculture, livestock farming, or the poultry industry. Wheat, corn, hemp, millet, sorghum, and peanuts were the regular feeder fare, although we need to appreciate that suet and animal fat had always been typical winter offerings. Sunflower seed was part of the mix, but its popularity was based largely on availability, and that grew steadily following the Second World War.

The story of the development of sunflowers from a plant cultivated for millennia on the American Great Plains, through export to Europe in the 1500s, its utilization by Russian peasants from the 1860s, the return of the plant to North America via Russian Mennonites shortly thereafter, and the remarkable scientific research in both the United States and the Soviet Union deserves an entire book in itself (although it is nicely summarized in *Feeding Wild Birds in America*).[63] For our present purposes, however, we can cut to the key events that resulted in the growth of the sunflower industry that led to the availability of sunflower seeds for feeding birds. In short, it's all about oil.

The indigenous peoples of the Great Plains had exploited the rich although tiny seeds of the native sunflower plants for millennia, using them for food, oil, medicine, and ceremony. Crucially, these insightful people engaged in remarkable selective breeding, eventually increasing the size of the seeds 1000-fold.[64] These were the plants "discovered" by the Spanish in the 1500s. Soon they were widely grown in suitable areas of Europe. Having evolved in a hard, arid environment, the plant was naturally drought resistant, a feature that certainly saved many lives among the Russian people who cultivated it extensively throughout the harsh interior of their country. The relatively high edible oil content of the plant's seeds attracted the attention of the Soviet agronomist V. S. Pustovoit (1886–1972). Between 1924 and the 1960s, through careful breeding experiments, Pustovoit was able to increase the oil content from approximately 20% to 40%, an outstanding and unprecedented achievement. In terms of improving the efficiency of a crop—the outputs of useable material compared to the inputs of water and fertilizer—these were spectacular results. For advancing the welfare of the Soviet people Pustovoit was awarded the Order of Lenin and the Red Banner of Labor.[65]

These developments were being noted from afar, especially in the oilseed industry back in the United States. Despite the Cold War climate of the 1960s, the agricultural company Cargill was granted permission for their agronomist Dick Baldwin to visit several research stations in the Soviet Union to learn more about these important developments. During this 1966 tour, Baldwin met Pustovoit (then aged eighty) and having seen the much anticipated 40% oilseed, he was dumbfounded to be shown a package containing the latest batch of sunflower seeds with a new benchmark: 46% oil. The Soviets may have been winning the Space Race at the time, but for Baldwin, this was even more important. The story goes that Baldwin asked if he might keep the package of precious all-black sunflower seeds.[66] Knowing that such a request would most certainly be declined by the minister for agriculture, his Soviet host handed the seed to Baldwin's local interpreter to be enjoyed as a snack. We can only imagine Baldwin's silent torment as he watched the priceless specimens being consumed before his eyes, and then his bewilderment when the interpreter quietly handed him the remaining 100-odd seeds during the car trip to the next stop on the itinerary. Those now legendary Soviet sunflowers somehow made it back to the Cargill base in Fargo, North Dakota, where they formed the genetic basis of a huge increase in oilseed production during the 1970s. Despite the apparent potential for starting an international incident during nervous times, the event appears to have fostered a rare and beneficial collaboration between US and Soviet scientists, with many subsequent trade and research visits. This region in eastern Europe remains at the heart of the global sunflower supply: in 2012, sunflower production in Ukraine and Russia together accounted for 44% of the 37 million tons produced worldwide, with the United States providing 1.26 million tons.[67]

Hence, the significance of the black sunflower to our story. Although developed primarily for high levels of edible oils, compared to the traditional striped variety, black sunflowers have larger, "meatier" kernels, more protein, and, most important, much thinner husks. In other words, this was a bonanza of high-protein, easily opened seeds for a suite of the smaller species—titmice, chickadees, finches—while all the larger species also found them even easier to consume. The arrival of these sleek black seeds during the 1970s was most certainly a key event in the history of bird feeding.[68] And that is a claim best backed up by the birds themselves: they came in droves, providing the best possible reason for the consumer (this time, the humans with money) to return eagerly to the stores for more of the same.

Black sunflowers were not simply favored by birds at feeders. My marginal role in the saga of this seed was as a young researcher attempting to understand why so many birds were flocking to what was a relatively new crop for the dry interior of eastern Australia. This was the early 1980s and my first real job since graduating with a degree in wildlife management; it was to be a transformative experience. To see flocks of over 10,000 parrots (mainly huge Sulphur-crested Cockatoos and Galahs as well as Cockatiels) and pigeons, finches, and sparrows descending on those vivid yellow fields was an astonishing sight. Why were they concentrating on sunflowers when other feeding resources were abundant nearby? My eventual, somewhat useless conclusion, was that nothing—including all their natural foods—was as attractive to a dozen species of granivorous birds as a field of almost ripe black sunflowers.[69] This was not particularly welcome news for the Oilseeds Board, who had funded the work, or the farmers growing the crop. Nor would it be the last time I failed to come up with a solution to a problem. But, mysteriously, it did set me on a path toward behavioral ecology: understanding how the behavior of animals is influenced by their environment. And food is always fundamental to this.

So the advent of black sunflowers was a major event. Around the same time, an entirely new type of seed appeared on the scene, niger (*Guizotia abyssinica*), another oil-rich grain used traditionally as cooking oil in eastern Africa and southern Asia. In the United States, this seed was renamed "Nyjer" in 1998[70]—ostensibly to clarify its pronunciation—and registered as a trademark of the Wild Bird Feeding Industry. It is also, inaccurately, known as "thistle." This seed was much smaller and lighter than most of the other seeds available during the 1960s, but its arrival coincided nicely with the lift in experimentation in feeder design. Very quickly, nyger-specific feeders became available, including small stick-on window units and purpose-designed tube feeders with appropriately smaller access holes. The new seed was, like the black sunflower, almost instantly successful, and for essentially the same reasons: this new food attracted a different group of species. This included Pine Siskins, Goldfinches, Dunnocks, and redpolls, species with strong but slender bills that can extract the seeds while the sparrows and larger species cannot.

The Choice

The message was becoming steadily clearer: to attract more species, deploy a wider variety of foods in as many ways as possible. Clearly, this was a market opening up to more choice, and the number and variety of seed mixes began to explode. But so did the cheap options. Modestly priced bags of attractively colorful blends of exotic-looking grain appeared in every shopping center and garden store, and were purchased in great amounts by the growing numbers of people interested in feeding and watching birds at home. We have all been there at some stage I'm sure; a big cheap bag of oats, rock-hard sorghum (milo), cracked corn, wheat, and what looks a lot like small stones. Apart from some House Sparrows and a desperate feral pigeon, many species that arrive to inspect a feeder of this stuff typically leave after little more than a taste. It's both disappointing and almost belittling to be rejected by your own birds. Although this experience is all too familiar to feeders today, general knowledge concerning the preferences of species for the various foods was hard to come by and often limited to specific areas. Most people simply purchased whatever was available locally and hoped for the best. Often this resulted in the feeding of species most people don't really want to encourage: sparrows, starlings, grackles, magpies, and the like. At a time when the variety of foods and feeders available allowed people to target particular species, what was needed was reliable information on which species liked what.

While other preference studies (including by the British Trust for Ornithology)[71] had been undertaken around this time in the late 1970s, the most significant were a series conducted by Dr. Aelred Geis for the US Fish and Wildlife Service. Starting with the study undertaken in Maryland entitled *Relative Attractiveness of Different Foods at Wild Bird Feeders*[72] (published in 1980), this research report was influential well beyond its limited geographical scope, generating considerable debate and a plethora of similar trials. During 1980–1985 Geis continued to test seed preferences, arranging volunteers to record the comings and goings of birds to carefully designed feeders in Maine, Ohio, and California, as well as Maryland. This was pioneering "citizen science" on a fairly grand scale, long before the concept was first formalized. Geis was able to base his findings on solid data with his observers logging over 700,000 visits. Although his results would be regarded as fairly obvious by today's collective knowledge, these were the first preference studies actually based on sound choice experimental methodology. Using two seeds already known to be attractive to a lot of species—black-striped sunflower and white provo millet—as the standard, Geis tried fifteen different types of seed and an additional six variants of sunflower, presented to his discriminating clients (the birds) in randomly positioned lots of four. As well as the obvious—wheat, corn, milo, and oats were distinctly less popular to most species—Geis demonstrated that many species preferred the same seeds in different parts of the country. He also confirmed the all-round attractiveness of black sunflowers but concluded that the best strategy was to mix it up: plenty of variety and a mixture of presentations, platforms, hangers, and straight onto the ground. The number of seed companies was burgeoning, and those paying close attention to this research were soon modifying their offerings. Many major marketing and production decisions and the composition of seed blends developed during the 1980s resulted from the Geis work.

Proliferation

The next marker in this story is not the appearance of any specific item or event but the spectacular proliferation of products that accelerated during the late 1980s and early 1990s. This period may be regarded as the advent of the wild bird feeding industry per se, with the founding or consolidation of a series of specialist companies providing a range of items for sale to a growing community of feeders and bird watchers now interested in feeding and willing to pay. This was a significant moment: when sufficient numbers of the public, almost certainly already engaged in regular feeding, began to seek to enhance their practice with commercial products. For reasons that have yet to be discerned, enough people were willing to buy the newly available seed packages and dispensers that a few visionary entrepreneurs decided to fill the demand.[73]

Although there are numerous individuals who could be mentioned, two stand out. In 1981, Jim Carpenter founded Wild Birds Unlimited, selling sunflower seeds and simple tray feeders from a store in Indianapolis, Indiana. Today, his company oversees almost 290 stores throughout the United States and Canada, selling more than sixty types of seeds for wild birds (including the recent addition of Jim's Bird-aceous Bark Butter) and a seemingly endless variety of feeders.[74] In 1987, Chris Whittles launched the first company in Britain to specialize in wild bird foods. Now known as CJ Wildlife, this is still one of the largest suppliers in the region and today has spread into nine European counties. The emphasis has continued to be on quality of the source products and the research basis of the many seed mixes and cakes. At last count, CJs offered 117 varieties of bird food and 184 different types of feeders and accessories.[75] Recall that in the 1970s, relatively few people purchased the mere handful of products that were then available. From free crumbs to the Deluxe Evri-Bird Crystal Dome Multi-Feeder in a few decades: a remarkable and largely unexpected industry!

A World of Feeding Birds?

This vibrant, innovative, and undoubtedly lucrative industry is still relatively new. It is also unashamedly ambitious, with all sorts of plans to expand the offerings available: new niche seed mixes, high protein "fat plus" balls, ever more elaborate dispensers, the latest technology aimed at thwarting the fluffy-tailed marauders, and so on. The problem for the companies involved is that this particular sector seems to be approaching saturation. Recruiting new participants has always been a challenge, and while the many imaginative advertising and promotional programs still seem to be working in the big established markets of North America and Britain, some companies are starting to think more broadly: the obvious next step is new markets—in new countries—altogether. But this is not as straightforward as it seems. It may not be as simple as translating the packaging and replacing the bird image used in the advertising. Local practices and attitudes can have a major bearing on the success of a new product attempting to make a mark in a different country. One example I am aware of relates to attempts by a major seed company to break into the Australian wild bird food market. While Australia is a country with a large proportion of people feeding, the majority of foods being offered is not packaged seed. Every Australian supermarket does offer various wild birdseed products, but these are not big sellers. What a lot of people are buying for their birds is made of high protein (i.e., meats and cheeses) for the various carnivorous species that are the typical feeder birds of Australia. The natural diet of these species is meat in the form of worms and grubs. If so, perhaps it would be possible to manufacture something like a high-protein artificial worm. This was a concept that got as far as a trial product—they looked something like segments of whole-wheat spaghetti and just might have worked (I have seen Australian Magpies stealing pasta from alfresco cafés), but the project was suddenly canned for reasons that were never revealed. (Of course, live, dried, and frozen mealworms are now readily available in many countries.)

Ideally, the seed companies need places where lots of people already feed birds regularly and may, therefore, be inclined to buy items to promote their well-being. As we have already discussed, people everywhere have always fed birds informally, without the need for purchasing anything. However, certain communities and maybe even entire societies are more inclined to feed than others, and to do so in an ordered, systematic fashion: organized bird feeding is a regular feature of some countries but not of others. Being aware of these differences would be of great interest to the feeding industry, and it is highly likely that they already have detailed information on the practices of many potential markets. This is exactly the sort of data I would love to see, but it is unlikely that I am going to gain access to such "intelligence."

To the best of my knowledge, nobody has even attempted to determine the feeding status of entire countries; no such information appears to be available in the public domain. There are, however, some excellent recent investigations of the characteristics of feeders in general from several parts of the world, which include the fascinating issue of what motivates people to feed birds. This complex topic is so central to our story that it is explored in a section by itself (Chapter 7), but it is peripheral to what I am proposing here. I am considering feeding at a much larger scale: the apparent national, possibly even cultural, tendency of a large proportion of the population to practice wild bird feeding. And no, I'm not going to specify a particular number: "large" will have to do.

Before starting, we need to return to the difference between simple spontaneous feeding and the more organized, planned practice. This is important as it allows us to distinguish between countries where feeding is common throughout and countries where it is not, without the distraction of the universal "tossing a chip to the gulls" version of feeding. This is part of our grand plan of trying to piece together the influences and circumstances that have led to the international phenomenon and industry we see today.

The usual approach when investigating a new research question is, of course, to consult the literature. After more than a decade of an unhealthy obsession with this topic, I am fairly sure I have seen almost everything published, at least in English, and have a decent network of colleagues to interrogate and query, including many in non-English-speaking countries. After consulting all these sources, I feel confident in stating that no one has yet attempted a comprehensive bird feeding survey of the countries of the world. And I am not about to try. What I will attempt, instead, is to list those countries where organized bird feeding is a common practice. What I will not be attempting is to list countries where feeding does not occur; this would assume that such information was actually available. It is not, so the absence of a country from the list means only that I could find no relevant information.

A separate but important issue is those countries where organized bird feeding is practiced but on a more limited scale, perhaps restricted to specific societal groups or targeting certain types of birds. Initially I attempted to compile such a list but was forced to abandon this task simply because of the paucity of data. We simply do not know what is happening in most countries. Instead, I will discuss a few places where we know at least something.

The list of feeding countries is based on what I could glean from the published material and many discussions with international colleagues. To these worthy and studious endeavors, I must add a far more enjoyable mode of research. In late 2014, I attended a large scientific conference in Malmö in southern Sweden, covering a topic nothing whatsoever to do with bird feeding. Finding myself surrounded by colleagues from almost every European country as well as a wide selection of others, how could I resist? Over many a coffee or beer I would innocently change the topic of conversation ("So, do they feed birds where you come from?") among a group of, for example, South Africans, Belgians, Brazilians, Japanese, Poles, and Swedes. This unorthodox—and unauthorized—approach turned out to be illuminating. My initial queries started extensive e-mail exchanges and debates, many of which continued for months. Obviously my approach cannot be regarded as a legitimate method of data collection, but it did corroborate information I had already gleaned.

What this preliminary and incomplete approach allows me to do is list countries where feeding is commonly practiced. To aid clarity, the countries are listed in continental groups.

Countries where organized wild bird feeding is practiced by significant proportions of the population

Europe: Austria, Belgium, Czech Republic, Denmark, Estonia, Finland, Germany, Hungary, Iceland, Ireland, Latvia, the Netherlands, Norway, Poland, Romania, Sweden, Switzerland, United Kingdom

North America: Canada, United States

Oceania: Australia, New Zealand

As a scientist trained to assess the strength of arguments based on the information presented, the methods used to collect it, and the results of the statistical tests applied to the data, this list is not something I am willing to die for. Nonetheless, I feel quietly confident about the countries included, although it is undoubtedly incomplete. There will be others I have missed, but those countries listed have been corroborated from several sources.

The first feature to note about the countries listed is that of cultural heritage: all are either European or societies of predominantly European background. Although most are from the Northern Hemisphere, the two Southern Hemisphere countries, Australia and New Zealand, were settled relatively recently by Europeans (South Africa is a special case discussed below). This might tempt us to conclude that wild bird feeding is simply "European." A closer inspection of the list, however, also reveals that most of the European countries listed are from the northern part of the region while the European countries not listed are all from the Mediterranean. France, Spain, Portugal, Italy, and Greece are all missing. So, to come to an early and oversimplistic conclusion already, countries where bird feeding is widely practiced are typically colder. Noting the two Antipodean exceptions, inclusion in this list is by climate: a national tendency to feed wild birds seems to be fundamentally about provisioning birds in winter. It would appear that some individual practices are reflected at the national level.

Such revelations may be embarrassingly self-evident, but bear with me; there are important patterns yet to emerge. The next task is to consider those countries not listed but where some level of organized wild bird feeding certainly occurs but is practiced by a relatively small proportion of the national population. The group below is very obviously just a start but may be useful for discussion:

> Argentina, Brazil, Costa Rica, France, Greece, Italy, Mexico, Peru, Portugal, South Africa, Spain

We can start with the Mediterranean countries already mentioned. There really does not seem to have been a tradition of widespread organized wild bird feeding in these places, even though some regions—the north of France, the Pyrenees, the mountainous parts of Italy, for instance—certainly experience sufficiently cold winters to at least provide the normal setting for the most familiar expression of the practice. Some such feeding almost certainly does occur in these places, but I can find no information of how much. Evidence is accumulating of genuine organized, regular feeding becoming popular in some areas, usually because of promotion by bird and conservation groups. Indeed, there are some localized sites where bird feeding is relatively intense, but due almost entirely to the presence of large numbers of northern Europeans. This is particularly evident in Spain where enclaves of British (Costa del Sol) and German (Empuriabrava) retirees now live, bringing their feeding habits with them. Similarly, cultural influences appear to be associated with feeding being far more prevalent in British Gibraltar as compared to the adjacent Spanish townships.

It has proved difficult to find much in the way of useful information for most Central and South American countries, although there can be no doubt that wild bird feeding occurs to some extent in most. This does seem largely focused on hummingbirds and occasionally certain parrot species. Given the European influence mentioned already, it is possible that the practice may be restricted to certain social groups. But it is difficult to be any more explicit.

In South Africa, a sophisticated feeding and wildlife gardening culture has developed recently.[76] In terms of the ready availability of feeding-related products, the detailed nature of the advice, and the enthusiasm of the participants, the scene in South Africa is clearly vibrant. Nonetheless, the practice is distinctly confined to the minority European community and evident mainly in certain cities, notable Cape Town and Durban.

Winter Only?

Beyond the simple proportion of the population engaged in feeding, the critical difference between the "minority" feeding countries and those listed as "significant" feeders is the issue of winter feeding. As mentioned already, the feeding of birds in winter correlates directly with the widespread expression of the practice. In countries where winter feeding does not occur regularly, the proportion of the human population feeding birds is much lower. Furthermore, in the "feeding" countries, the feeding of birds at times other than winter has become a critical point of discussion.

This point was forcefully emphasized during my opportunistic conversations in Sweden with colleagues from many European countries. Indeed, it was somewhat surprising to hear the same strong sentiment expressed repeatedly: yes, bird feeding was common in many European countries but only during winter and certainly not at other times. Vigorous debate then ensued as to what might be the most appropriate triggers for the "proper" starting time for provisioning—a particular date, the first snow, the first frost, for example, were all advocated—and similarly, when to stop. These discussions often became quite animated when it became clear that some countries actually practiced feeding all year round. For some, this was almost a violation of a well-established, cross-cultural norm; for others, continuous feeding was a simple and natural extension of the winter practice. The different perceptions and reactions among these well-educated and well-traveled colleagues was fascinating to witness and indicated that this issue was of great significance to the story.

All-season, indeed continuous, feeding, appears to be such a profound change in long-held practices, if not traditions, that I consider this to be the most significant development in the recent history of wild bird feeding. Exploring how this change came about and why it is important requires a chapter of its own.

3

The Big Change

Winter or Always?

It is a glorious golden autumn day, something fairly unusual for central Poland during September and definitely to be enjoyed while it lasts; the wind, rain and cold will be coming any day now. Skeins of migrating Graylag and Bar-headed Geese steadily cross the clear blue skies, while the hedgerows are alive with southbound warblers and flycatchers. My guide and colleague, Piotr Tryjanowski, a professor at nearby Poznan University, is striding through the fallow fields toward a huge cylindrical roll of hay, apparently having spotted something of interest. I am doing my best to keep up, but the "new" birds keep appearing: Whinchat, Crane and just now a magnificent Black Woodpecker—Europe's largest—has flown by, direct but heavily and unmistakable. "There, and there too!" Piotr whispers hoarsely, peering in one direction with his binoculars and pointing his left arm in another. Somehow I manage to follow his double directives and spot them both in succession: a Great Grey Shrike *and* a Red-backed Shrike! Five "lifers" in the space of half an hour; I am ecstatic! Piotr shrugs. "Good but not so special. We expect them at this time of the year, but I guess you are lucky."

The Great Grey Shrike is just one of the species Piotr has studied intently, but it is certainly a favorite. "Unfortunately, you rarely see shrikes in gardens," he explains, a pity because Piotr has just added wild bird feeding to his lengthy array of research interests, and it is the reason for my visit. "There is a long tradition of feeding birds in Poland, especially the hanging of fat and suet on trees. This may have started when farmers noticed birds feeding on hanging pig skins, though today most people probably use commercial products." Feeding mixtures are widely available but most people seem to use the "little house" feeders on a pole. Feeding is also a prominent community activity, with many informal groups of neighbors organizing a roster to keep a local communal feeder, located in a park or street, supplied with seed. Probably a majority of schools in Poland maintain a feeder in the school grounds. "It's just normal here," Piotr observed.

Among many things, Piotr is interested in the differences between species, sexes, and even individuals in their eagerness to approach feeders. "I suspect we will discover that particular birds have personality traits which mean that they are more likely to take risks and be first at the feeder. If these birds breed better, survive longer, and pass their traits on to the next generation, it could mean that feeding is producing a 'feeder personality': birds that dominate in gardens and keep out the others. But we need to do a lot of careful research first." This was but one of many dimensions of future research into the implications of wild bird feeding I learned from Piotr. As we discussed these questions over a customary late-night beverage, Piotr's new research program seemed a natural extension of a common and important traditional practice for the Poles. Although a few new products may have been added recently (new seed mixes, different feeder designs), this has been a long-established activity in Poland, and something I was keen to experience for myself. Unfortunately my visit was just too early. "Sorry, my friend," consoled my host. "Feeding starts usually on the first day of December around here, maybe a little earlier if we get a lot of snow." In Poland, many traditions are changing, but not this one. "It's only in winter, only when it's really cold! Everyone knows that!"

Cold it is, a lot, in the Alpine village of Niederthai in the mountainous Tyrol region of Austria, the childhood home of Ann Göth, a friend and collaborator of mine. This wonderfully picturesque location has attracted tourists for hundreds of years, who come to ski or enjoy the dramatic scenery during the short summer. But visitors typically stay only briefly;

living permanently at 1600 meters (5200 feet) is challenging for both people and animals. Now living in a very different setting in Australia, Ann recalls her childhood experiences: "Bird feeding when I was young? Yes, it was certainly done, but more often in the towns and not so much in the rural villages in those days. Where I grew up we had snow for at least six months of the year, and when it was really cold we would see the birds gathering seed from the cattle food. I remember that feeding was mainly for the tourists visiting the chalets; they used those netted balls with fat and sunflower seeds in them. But there were also places where the hunters fed deer and wild boar in the winter, and the birds often got a good feed from those feeding stations. I loved to stand quietly near these places, just to watch the birds on many occasions, often scaring nearby skiers who thought I was a gnome from the woods."

Ann's friend and colleague Monika Rhodes grew up in a suburb of Esslingen in southern Germany, about 300 kilometers (190 miles) north of Ann's Austrian village and in a much more human-dominated landscape. During a recent get-together in Sydney, I was able to briefly interrupt their conversation to ask about their childhood experiences of bird feeding. Having listened to Ann's description of feeding in Austria, Monika was keen to add her perspective: "We had a wonderful birdhouse feeder built in the style of an Austrian mountain house (just like the real houses in Ann's village) where one could open the roof to fill the inside with seeds—sunflowers and other birdseeds. My father took great care to clean it and fill it up during the snowy winter months. We had it on our balcony and watched the birds feeding while we had breakfast or lunch in our dining room. I grew up with binoculars and a bird book next to our dinner table, so every time we saw a new bird we quickly tried to identify it.

"We ensured that we only fed the approved birdseeds recommended by our local bird watching group [*deutscher Vogelschutzbund*] where I was a junior member for many years," continued Monika. "We hardly ever fed the birds with the netted balls Ann describes (though sometimes did so in very cold long winters, usually for the tits). We were told that the fat was the wrong type and not good for birds and hence did not put these balls out. We never fed them for long periods as back then the recommendation was that the birds needed to look after themselves. So feeding was mainly during the snowy cold months when food was hard to get. However, I think my father secretly enjoyed having the wildlife around the house and may have fed them longer than needed. Still, as a young conservationist I always tried to educate my parents! Feeding birds in summer and from the hand was not done and rather frowned upon."

These observations, from the present and the past, may be simply personal and anecdotal, but they also illustrate the traditional view of wild bird feeding in Europe: in most places, it is still a practice conducted during the winter only. As described in the previous chapter, in a clear majority of countries where feeding is widely practiced, winter-only feeding is definitely the norm. It is a practice regulated by various mainly informal "rules" or expectations of local communities, some so entrenched that they probably qualify as national traditions, while there are also more formal guidelines from conservation and bird groups. I have emphasized this point perhaps too much, but it is important to appreciate that for a significant proportion of the feeding world, the provisioning of wild birds is, quite explicitly, winter feeding.

The realization, therefore, that a movement toward all-year-round feeding is rapidly gaining ground would certainly be disturbing to many, especially European, feeders. Although this has been occurring for some time in North America and the United Kingdom, for much of Europe such a fundamental change would seem unlikely or even unthinkable. Therefore, the popularity of a serious book advocating the complete abandonment of the winter-only tradition and instead promoting bird feeding throughout the year is, for some people, genuinely shocking.

Vögel füttern, aber richtig ("Feed Birds, but Correctly"), by Peter Berthold and Gabriele Mohr,[1] appeared in Germany in 2006 accompanied by an unprecedented level of publicity and comment for a bird book. As expected, there was concerted opposition, especially from within scientific circles, but generally strong support from feeders and bird watchers.[2] What made this publication stand out from the many look-alike gardening and hobby books was the authority of its writers and especially its audacity. The senior author was Professor Peter Berthold, a renowned ornithologist and specialist on bird migration, and he did not hide his mission. The book's subheading blatantly declared: "Feed all year!" This was clearly provocative, yet it struck a chord with the German public. Within the first two years the book sold more than 50,000 copies, with a second edition published in 2008, an extraordinary achievement. As such, this book deserves to be the most recent exhibit in our history of wild bird feeding presented in the previous chapter. However, given its apparent influence on the pattern of bird feeding in Germany and beyond, it is appropriate to feature this remarkable book—and its main author—in some detail (a little later).

Why the Change?

The growth and diversification of the wild bird–feeding industry in recent years has been staggering, and it continues to expand. This means, among other things, that more people are feeding and that these people are also feeding more frequently: more food, more often, and in some places, throughout the year. Of course, we do need to acknowledge that in many countries—primarily in warmer climates—bird feeding has always been a year-round practice. Think of the seemingly universal provisioning of hummingbirds right through the Americas, or the typical practice in Southern Hemisphere countries where feeding is entirely continuous. In these places, helping birds through winter has not been a major motivation, although assistance through other forms of "difficulty" such as severe droughts or in the aftermath of serious storms certainly occurs. This is not what I am exploring here.

What the "big change" refers to is the apparently widespread and permanent move to continuous feeding in those countries where winter-only feeding was long the norm. The potential implications of such a massive increase in food being provided for wild birds is central to our story, and we really do need to explore how this change in what was a traditional practice came about. To my knowledge, this aspect of the bird-feeding story has not been researched in any detail, so again, our discussion will inevitably be based on limited sources of data and heavily influenced by my discussions with several experts and key players. Let's take a closer look, then, at the three places where this change has been particularly prominent, and for which there is at least some information: North America, the United Kingdom, and Germany.

North America

I am again obliged to Baicich, Barker, and Henderson, the authors of the *Feeding Wild Birds* booklet and the more recent *Feeding Wild Birds in America*, for their historical description of the development of the feeding industry in North America.[3] Although similarly detailed information is simply not available for most other places, it seems likely that organized feeding practice had commenced earlier and was far more advanced as a commercial activity in the United States and Canada than anywhere else in the world. Yet while the number of seed companies and feeding products proliferated during the middle of the twentieth century (as described in Chapter 2), by far the majority of feeding up until the mid-1950s remained winter-only; "it was still viewed as a way to help save birds."[4] Perhaps significantly, there appear to have been far fewer of the rigid traditions among North American feeders more typical of winter feeding in Europe. The impression is of a more relaxed, informal hobby, with an emphasis on enjoying the presence of birds. Certainly, the bulk of the feeding activity was conducted during the cold months, but if folks wanted to keep feeding, few would have objected. This scenario is exemplified by the typically year-round feeding of hummingbirds. Everyone loves hummers, and the reward of attracting these living gems is such that plenty of people do so continuously. Indeed, it is quite possible that the presence of these charismatic birds throughout the year has influenced the move to all-year feeding in general.

The first commercial move away from the winter-only patterns seems to have been the arrival in 1956 of the explicitly labeled "Four Season" wild bird mix. This new product heralded the start of a somewhat novel concept: summer feeding. As described by Baicich and coauthors, "A new bird feeding niche had been created."[5] This so-called niche was that of continuous feeding, a practice "both entertaining and practical," a phrase nicely encapsulating the enjoyment of the people involved and the assumed benefit to the birds. At the same time, there can be no doubt at all that this niche represented a gigantic commercial opportunity for the companies ready to respond.

Put simply, the move to all-year feeding in North America seemed aligned with the public's willingness to try new products, rather than needing to be convinced by careful arguments in support of such a change. For this particular region and society, the change now seems almost inevitable. It also seems to have been generally free of controversy and even scientific discussion.

But please don't assume that the seasonal and geographical pattern of bird feeding is uniform across the whole of North America. Although recent surveys comparing regional differences in feeding practice confirmed that large proportions of the population fed birds, the highest participation rates were associated with the northern coastal states, east and west, of the United States, with the lowest in the central zone.[6] There appears to be very little published on when people feed, but my impression, derived from numerous North American contacts, is that most of the people who feed birds do so throughout the winter but that participation is slightly lower at other times, down to perhaps 70% of feeders.[7] This is still a strong majority feeding all the time.

United Kingdom

As in many other European countries, wild bird feeding in the UK has a long history, although the timing of the modern form of a feeding industry appears to have paralleled the development occurring in North America. Various traditions and informal rules associated with when to start and stop feeding seemed to have applied, with innumerable variants (for example, in places in Ireland, winter feeding started with the first two successive frosts, or the sighting of "frolicking" hedgehogs), and it appeared that, until fairly recently, the overwhelming majority of feeders in Britain did so only in winter.[8]

The change to year-round feeding in the UK appears to have been under way informally for some time but with a definite acceleration during the late 1980s and thereafter. Assuming that the typical practice was mainly winter feeding until at least the 1950s, the first British survey (the often-cited Cardiff research)[9], which was conducted in the early 1980s, reported that about 40% of these households were already feeding in summer, surely an unexpectedly high proportion for those times. By the early 2000s, however, a large independent survey of five major English cities reported that 64% of respondents fed birds "at least weekly"[10] (which implies regularly and continuously), while a 2013 study focused on the city of Reading found a virtually identical rate (65%) were feeding birds year-round.[11] Clearly, the traditional "norm" of feeding only in winter has become a practice of a minority.

And this is where things get really interesting, for me at least. While it appears that small but increasing numbers of British feeders were starting to feed throughout the year during the 1980s, during that period the two most respected ornithological institutions in the UK—the Royal Society for the Protection of Birds (RSPB) and the British Trust for Ornithology (BTO)—were both clearly and implacably opposed to feeding outside winter. For example, in official advice provided to its huge membership at the time, the RSPB indicated that there was "no need to feed birds between April and September" because natural food should be abundant during those months.[12] Furthermore, it was particularly important that artificial foods not be provided at these times as birds "may be tempted by easy food that can choke their young." This was also the unqualified position adopted by the BTO, which recommended that feeding be limited to the period of August to mid-May, stopping then to avoid the period when tits had nestlings.[13] From both these leading groups, the message was clear and straightforward: Don't feed outside winter.

Within a decade, however, the "official" message had changed completely. By the end of the 1990s, although both the BTO and RSPB continued to promote winter feeding as always, the formerly solid stance against feeding at other times had disappeared entirely. Instead, there were distinctly less negative, seemingly more considered statements, unequivocal in their support for year-round feeding. For example, the BTO:

> The modern approach to garden bird feeding is to use a range of foods that support the specific nutritional requirements of a wide range of species over the course of the year. There is scientific evidence highlighting the positive effects that the provision of supplementary food can have on birds.[14]

And the RSPB:

> Although winter feeding benefits birds most, food shortages can occur at any time of the year. By feeding the birds year round, you give them a better chance to survive the period of food shortage whenever they may occur.[15]

These statements may seem so sensible, maybe even so obvious to us today, that you may wonder why I bother highlighting them at all. My point is to draw attention to the magnitude of the change in stance evident here. This has not been a subtle shift in attitude; it's a complete 180-degree reversal. One might wonder how this profound switch came about. After all, both the BTO and the RSPB are highly respected organizations, with huge memberships, powerful political clout, and proven records of successful conservation leadership. People (even politicians) listen to them and take note: a change in policy or direction would certainly be influential. This is because, crucially, both are strongly science-based groups, conducting world-class research and providing vital advice to governments and agencies throughout the United Kingdom and beyond. If these people made such a big change on such an important issue, surely it was going to be based on intensive research and solid evidence.

For someone with my interests, it was therefore imperative that I find out as much as possible about this crucial period—the 1980s and 1990s—of major change in the practice of wild bird feeding in one of the key countries in this story. Although I had assembled all the materials I could find, there was little in the way of background from this era. Both the BTO and RSPB have been engaged in ornithological and conservation research for decades and are committed to ensuring that their publications are easily accessible. Hours of searching their websites yielded lots of reports and documents on a diverse selection of topics, but remarkably few on feeding. One obvious problem may simply have been the lapse in time; these events took place decades ago. Most of the people involved had probably retired; maybe the key papers and reports I sought simply predated digitization (and therefore were not available online). Possibly relying on the Internet was actually not the best method in my quest for clarity. I actually needed to talk to people, preferably face to face. Although there was any number of relevant issues to discuss with people within these organizations, for me the most intriguing question was: What led them to change the stance on feeding practice? What better excuse to visit these two iconic British institutions in person. It took some organizing, but in November 2014, that is just what I did.

The Lodge For some naive reason, I had expected the headquarters of the Royal Society for the Protection of Birds, founded in 1889, and Britain's largest conservation charity, to be based in either an august Georgian building in central London or, in keeping with their recent contemporary rebranding (new motto: "Giving nature a home"), a glass tower somewhere like ultratrendy Canary Wharf. But the RSPB HQ needs to accommodate hundreds of staff, and that requires plenty of space. So for over 50 years, its primary center (there are also numerous regional offices) has been in the countryside of Bedfordshire, about 50 miles (80 kilometers) north of

London. The Lodge, an old, grand, and rambling red-brick complex, is situated deep within the Lodge Reserve, one of more than 200 nature reserves managed by the society. The reserve is visited by huge numbers of birders, even on dull, rainy days such as when I visited in November.

While working on this book I have e-mailed, phoned, and Skyped with dozens of people from all over the world, but meeting them in person whenever possible always seems more comfortable. Many of these people didn't necessarily know me that well and may have been at least a little wary of my agenda. Meeting face to face proved crucial in the development of this strand of the story. I tried to convey the essence of my mission as sincerely as possible and attempted to articulate my desire to understand. I think it worked; my query about the basis for the change in seasonal practice led in every case to a thoughtful pause, followed by a range of responses that might be summarized as: "That is an excellent question," "I really don't know but someone will," and "I'm sure there is a report on that somewhere . . ." The reality of the time lag involved did prove to be an issue—these events really did seem to be somewhat historical and, given the more immediate concerns facing so many British birds, perhaps a little academic. It seemed a fair point.

I did find, however, real interest in this question among some of the RSPBs veteran researchers. Will Peach and David Gibbons generously discussed a wide range of feeding issues, including attempts to revive the faltering numbers of House Sparrows and recent research into the importance of grain supplies for many rural bird species (this work features later). As to the change away from winter-only feeding, both agreed it was a crucial though historical question, but could not pinpoint a particular study, event, or policy associated with the switch in the RSPB's stance. They suspected that the BTO had conducted research into aspects of this issue. Finally, and independently, both Will and David said with almost identical wording: "Support for summer feeding? That sounds just like something Chris Mead might well have had a hand in!" I noted the name immediately: Chris Mead had been a popular and very public figure within the companion organization, the BTO, for over 40 years, and his name had cropped up in many conversations already. As I was leaving the Lodge, Will offered one additional suggestion: "I suspect that you will also find that a certain Chris Whittles is going to play an important role in your story as well!"

As well as getting to know some of the researchers I had been reading for years, seeing the Lodge, and learning about the RSPB from within, this visit provided some tantalizing leads: a possible research report, and two very significant names. Chris Mead had been a legendary figure in British ornithology, a formidable combatant in the field of bird conservation, a bird ringer (bander) of international renown (having reportedly mist-netted over 400,000 birds), and most significantly, a well-known and respected spokesman for the BTO until his death in 2003, aged only 62. Chris Whittles, on the other hand, although recently retired, was still active. The founder of the huge and influential CJ Wildbird Foods company (described earlier), Chris was of a similar vintage to the other Chris; I wondered whether there might be a connection.

The Nunnery My next stop was unexpectedly close. The two organizations I was visiting, though entirely independent and distinct philosophically, are located only 56 miles (90 kilometers) apart in the middle of rural England. The BTO, founded in 1933 at Oxford, is now housed in a delightful converted nunnery (conveniently known as the Nunnery) on the edge of the Norfolk village of Thetford, an easy 90 minutes' drive up the A14 from the RSPB. Despite what I knew to be frantically busy schedules, several key people agreed to meet with me to talk about feeding. Kate Risely was the coordinator of the BTO's Garden BirdWatch, one of the largest citizen-science projects in the world, and an invaluable source of detailed data on garden birds and, of course, bird feeding. Using this massive collection of information to answer key questions was one of the tasks being pursued by Kate Plummer. This Kate had recently completed some of the most important experimental studies of the effects of diet on breeding outcome in birds[16] (discussed further in a later chapter). For my project, Kate Risely and Kate Plummer were an ideal source of information and opinion, and we ended up talking for hours. This continued as a prolonged discussion when we convened at the Black Horse pub that evening, joined by a number of other BTO staff. The issue of the BTO's change in stance on when to feed was especially fertile ground, although there was little in the way of concrete evidence that anyone could recall. Several people felt they could find something of value or interest, and Kate Risely declared that she would contact some of the individuals she knew from the era in question. Thoroughly invigorated by the debate,

I finally headed back to my country hotel, quietly anticipating further developments.

I spent the following day in the Chris Mead Library at the Nunnery, looking through every issue of the dully titled but fact-packed *BTO News Magazine*. As this excellent source of news and views had started in 1964 and continues today, it seemed a logical place to search for evidence of changing practices. Although various birdseed companies were advertising from the earliest editions of the magazine, and there were plenty of "how-to" articles on bird feeding and wildlife gardening, I could find no mention of nonwinter feeding. While I was engaged in this task, I was pleased to meet the magazine's editor through much of its life, Derek Toomer, who just happened to be passing through. When asked about whether he could recall any articles or editorials or anything at all, he suggested that I was unlikely to find a specific statement or event to mark the change; in his opinion, it has been under way since the late 1990s. Derek, who had been closely engaged in monitoring the trends and fashions among feeders, also suggested that it was likely that summer feeding grew indirectly as a result of the increasing availability of feeding products and new types of seeds. "I would say that the arrival of the seed companies, especially CJs, was significant, as was their involvement with both BTO and RSPB."

Meanwhile, Kate Risely's overnight networking efforts had yielded some fascinating—and familiar—opinions from several people involved with BTO during the period of interest: "I seem to remember it only as a pronouncement by Chris Mead. There might be a press release or *BTO News* article that would give you a clue." "Probably it was something that Chris Mead came up with, but I don't know the specifics. But if Chris Mead said it, then . . ." "My recollection is that we were prompted by bird food entrepreneurs to question the previous wisdom that summer feeding was a bad idea, at about the time when bird food started to become big business."

I may not be much of a detective, but even I could detect a pattern emerging here: an influential BTO personality and a "bird food entrepreneur." Sadly, the larger than life Chris Mead was no longer available, but Chris Whittles certainly was; it was time for a visit to Shropshire.

There was one more exhibit to be considered, however. A few days after my visit to the Nunnery, I received an e-mail from Kate Risely. Attached to the message was a BTO internal report by Patrick Thompson dated October 1987. "Looks like it could be useful," Kate added tentatively. It was titled *The Seasonal Use of Gardens by Birds with Special Reference to Supplementary Feeding.*[17]

I have to admit, I was not expecting much from this rather dated report. I already had found several studies from this period that were really just food trials, comparing preferences for different types of seeds; this was probably going to be something similar, I thought. Besides, my expectation was that surely the change in position by the BTO and RSPB had been linked to significant research of some kind. Many of the people I had spoken to felt the same, yet no one seemed to remember anything specific. This apparently forgotten report, which I had never seen cited, did not seem promising. It was late one night before I got around to actually reading it. And let me admit that my perception was just plain wrong. This work may have been overlooked, but it was exactly the kind of critical and focused research I had assumed must have been undertaken. And this was back in 1987.

The Thompson study was built on an existing (and pioneering) citizen science program—the Garden Bird Feeding Survey—instigated by the BTO 17 years earlier in 1970. (Interestingly, the report was produced for a company, Pedigree Petfoods, though their role is not described.) A dedicated group of about 200 people had been recruited for the survey to systematically record the birds visiting their feeders. It goes without saying that this was during winter; as we have seen, the official position was adamantly "winter only" at the time. Everyone was aware that feeder food was inappropriate for nestlings. Nonetheless, the BTO sent out letters to all the participants, inviting them to extend their feeding into the spring breeding season, a rather radical suggestion. Indeed, while the researchers were able to sign up over one hundred participants for the new part of the study, Pat Thompson also records that they received a number of letters expressing serious misgivings as well as several "seasoned participants who refused outright," stating that feeding nuts and other starchy foods had "been proven to be bad for nestlings." Receiving such feedback—faithfully reiterating the BTOs own advice—must have been at least a little perplexing.

Although this spring trial was conducted only once (I can't help but wonder how many participants kept it up afterward), it yielded a considerable amount of data showing that many species would indeed take advantage of feeders during spring. That was hardly a surprise, but it provided important confirmation that birds did use these additional resources. The crucial question, however, remained: Did the adults supply feeder food to their nestlings? The second part of the study involved an equally provocative proposition: provide peanuts for breeding tits and observe closely what happened, an experiment actually testing the central nightmare scenario of a nestling choking on a peanut. The ideal setting was readily available: Chris Mead (of course) just happened to have been ringing Great and Blue Tits near his home in Hertfordshire for over 10 years. To improve the chances of catching these birds, Chris had set up feeders offering peanuts next to his mist nets. Again, this feeding had been limited to the colder months, but for the sake of the study, he too agreed to keep feeding into the spring. (Is it possible that this was the critical moment for Chris Mead?)

Mead had already established a large number of nest boxes throughout the area, and his almost continuous presence nearby resulted in thoroughly habituated birds who could be observed from close quarters. Pat and his assistants patiently recorded hundreds of birds bringing food to their babies in the nest boxes, carefully noting what was on the menu. These valuable observations confirmed, despite the availability of peanuts nearby, that almost all—well over 90%—of the items delivered to the nestlings were instead insects. Some peanuts were, however, brought too—sometimes by Great Tits, extremely rarely by Blues—but no chicks succumbed as a result. While a very small number of nestlings were found to have died during the study, this was attributed to starvation, an unfortunate but entirely normal result noted in virtually every study of breeding birds. Crucially, none of these chicks died because of the peanuts. Finding sufficient insect food to raise a nest full of baby birds really does seem to be a major constraint on reproduction among tits. Feeding nestlings large items such as peanuts may be a sign of desperation on the part of the parent birds, and Pat Thompson's work suggested that this was likely to be extremely rare. Other studies have confirmed this finding and the fact that nestlings have on rare occasion choked on peanuts.[18]

The Thompson studies were indeed significant, providing important findings of direct relevance to the critical debate about the risks and potential benefits associated with nonwinter feeding. His report would surely have been influential at the time. It's just that no one seems to be able to recall his landmark research today.

Whittles' Wisdom The chance to meet Chris Whittles in person was something I had not really thought would be possible. Although I had met Chris briefly in 2010 at a symposium on bird feeding in London, there was no reason why he should remember me. Besides, he had since sold his share in the giant bird-food company he had founded, CJ Wildbird Foods, and, I had learned, was far from pleased with the direction the new management were taking the renamed CJ Wildlife. Why would he want to discuss any of this with a stranger? Would he be aggrieved when he learned that I wanted to visit the CJ factory itself (as described earlier), only a couple miles away from Chris's home in Shrewsbury, a place he had built from scratch but now refused to visit?

These concerns, of course, proved entirely unfounded. At my colleague Jim Reynolds's prompting, I contacted Chris myself. Whittles was well known for being forthright and opinionated; if he didn't agree, I would hear about it. I received an immediate response to my e-mail: "It will be delightful to see you again. Do come up to my home and leave plenty of time for talk—and lunch!" That seemed fairly positive. Jim and I drove up to Shropshire from Birmingham only a few days after I had been at the BTO, recent developments prominent in my mind. I also had no illusions about what I was about to experience: "The gospel according to Whittles!" as some were quick to suggest when they heard of my proposed visit. All agreed, nonetheless, his was an insider's view of the British and international bird-food industry that was simply unsurpassed. I knew it would be a biased view; I just didn't expect it to be so delightfully so.

Chris met us at the door of his spacious suburban home on the edge of Shrewsbury. After a firm handshake, we were ushered into a large lounge looking out onto a conservatory. In clear view through the glass panels was an expansive and well-maintained garden, festooned with bird feeders of numerous designs. Even though it rained steadily throughout the day, groups of tits, nuthatches, Blackbirds, and Woodpigeons continuously drifted by. It was an ideal if distracting setting for the conversation unfolding inside.

Chris Whittles, as he himself would readily agree, is not a man lacking in self-confidence or ambition. The personal characteristics that led him to become one of the undisputed champions of the international wild bird–feeding industry were soon evident: clear goals, decisive leadership, visionary risk-taking. But along with the entrepreneurial instincts was a sincere affinity for the birds. An agronomist by training, he had always been convinced of the necessity of a strong scientific and rigorous approach to business decision making. From its very earliest days in 1987, selling peanuts as bird food, the embryonic CJ Wildbird Food Company was characterized by an uncompromising attitude to the quality of its produce and a dedication to evidence. At the time, the typical wild bird products were low-quality mixes dominated by cereals such as wheat, striped sunflower seeds, and "rubbish" fat balls. CJs (according to its founder) revolutionized the industry in Britain by conducting some of the earliest preference tests where wild birds themselves were able to choose among a variety of offerings. A critical early breakthrough was the discovery of a clear preference for peanuts sourced from China over those from India or Argentina. Detailed nutritional analyses showed that the difference was in the fat content; the preferred nuts had much more oil. They were also more expensive to procure—and therefore sell—but Chris made the decision to go for the higher quality despite the additional cost. This commitment to highest-quality products was to become a defining characteristic of the CJs brand for many years to come.

Many other breakthroughs were to follow: the importation into the UK of the first black (high-oil-content) sunflowers in 1991; the development of the "ideal" table birdseed mix (less cereal, a variety of small seeds, and the important addition of kibbled peanut hearts); "peanut cake" ("the perfect winter fat source product"); and live and dried mealworms. The guiding philosophy was always that these items needed to be nutritious alternatives to the bird's natural foods. "Seeds, beech-mast, insects, and berries—these will always be the bird's first choice," but they can also be unreliable. "The substitution of well-balanced high-quality foods is the next best thing," Chris explained forcefully.

As the range of offerings expanded, they were all available via mail order. While CJs has at various times supplied products to supermarket giants such as Tesco and Sainsbury—as well as directly to BTO and RSPB outlets—overall the bulk of the business has been through mail order. That may seem fairly standard in these days of Amazon and eBay, but CJs was doing this in a big way decades earlier. That's brown wrapping paper, handwritten addresses, and individually licked stamps.

My Wi-Fi, Internet, Google-dependent mind was reeling. "But how did your customers know what was available?" I asked, naively. Chris responded with a knowing look. "Ah, our catalogs!" he said with a quizzical wink. "Would you like to see?" He disappeared into a nearby room and emerged with an armful of colorfully printed documents that he deposited on the table in front of me. "Every year since 1987. Have a look. You can see clearly how these evolved over the years." I picked up some of the older issues, more out of politeness, but as I began to flip through the pages, I was immediately engaged. These were far more than routine presentations of items for sale; right from the start they were glossy, full-color, and carefully designed. Obviously a wide range of products was featured, but the pages were also full of original artwork, gorgeous pictures of the birds that prospective consumers would have been interested in and attracted to. More than that, there were hints, bird-watching highlights, suggestions, and guidelines on how to assist the birds themselves. They were more like a magazine for bird lovers than a catalog of products. "I am very proud of these publications," Chris admitted. "I am convinced they were fundamental to our rapid success."

I had to agree. The immediate impression of the catalogs was of high design and production standards with clear rapport with other bird lovers. I picked up the beautiful 1994 issue, with a radiant inverted nuthatch on the cover, and flipped it open to the center pages. Under the heading "All about Feeding" I read: "For many years it has been recommended NOT TO FEED birds between April and July. The latest research shows this to be incorrect." A large chart displayed the twelve months of the year with horizontal bands of varying colors indicating when several common species—tits, Greenfinch, Chaffinch, Siskin—were most likely to visit British gardens. The message was sophisticated and convincing: the abundance of birds at feeders fluctuated in relation to their natural food sources, and suggested that they visited when they needed help, especially "in spring and summer when natural seeds are in short supply." The source of this particular information? "CJ Wildbird Foods test gardens, where all-year feeding has been carried out."

"You carried out your own research, and throughout the year?" I inquired. "Of course!" Chris stated bluntly. "It was necessary. I had to counter the misguided notion of feeding only in winter. Any observant feeder knows that the seed-eating species are hard pressed for much of spring into summer. They visit when they need to. But I had to convince people."

"People like the bird organizations?" I offered. Chris sighed. "It was about a 12-year battle—from the start of CJs in '87 to about 2000. They were adamantly opposed for a long time, both the BTO and the RSPB. But they eventually came around. Mind you, Chris Mead had always been on board; very supportive, despite his position in the BTO." (I didn't appreciate it at the time, but Mead had already been engaged in the "peanuts in spring" experiments with Pat Thompson by this stage.) "He would often join our presentations at the bird fairs and trade shows where he 'held court.' Great fun! Chris knew everything about birds and everyone loved him. And people listened to him. He provided a lot of good work for us especially in those early years. Look, that catalog you have there; turn back a few pages."

I did so. A two-page article titled "When to Feed Your Birds" explained that birds need help at different times of the year, not just when it is cold. "New research shows that feeding even in summer can be important." It was written by Chris Mead from the British Trust for Ornithology in 1994, long before any official change in attitude at the top. The "If Chris Mead said it, then . . ." theory was beginning to sound plausible.

"Mead was no fool," Whittles explained. "He only spoke up when he was convinced about something. The recovery of the Goldfinch is an important example." Having been a rare bird in gardens for decades, especially in southern Britain, numbers of this delightful and beloved bird have risen spectacularly since the late 1990s. According to Chris the reason was obvious: "Entirely due to our introduction of the new nyger seed. The goldies loved them! It was one of the real breakthroughs; stick up a nyger feeder and you got Goldfinch almost everywhere. It was a particular clear example of feeding having a strongly positive impact."

Our time with Chris Whittles concluded with a prolonged lunch at a delightful hotel dining room overlooking the swollen River Severn, a short drive from his home. The rain continued steadily and so did the conversation. Jim Reynolds (who later revealed how much he had learned during these discussions, from someone he had worked with for years) remarked that it seemed that some forceful personalities had a major influence on the evolution of the British bird-feeding industry. "You mean Mead and me?" offered Whittles. "Exactly! But let me add a couple more: Bill Oddie and Peter Berthold." We soon found out why. "Bill was *the* birding celebrity who fronted a key media campaign at a crucial time;

Peter was the famous German professor who single-handedly caused a people's revolution!"

In the UK, the case for year-round feeding during the crucial period of the 1980s and 1990s had been framed around the alarming news that the population sizes of many species of the smaller birds were declining dramatically. In a landscape utterly transformed by millennia of human occupation, and where traditional farmland features such as hedgerows and wetlands were being progressively removed by the intensification of agriculture, the unlikely habitats of British towns and cities became critically significant. The area covered by all the private gardens combined far exceeds that of all the national parks and conservation reserves combined.[19] These were areas where the decisions and practices of individual householders could have direct and positive impacts on the fortunes of a large number of bird species. Enhancing the habitat qualities by wildlife-friendly gardening—especially the provision of suitable foods—it was argued, was a simple yet effective way residents could make a difference. And this meant feeding beyond the traditional winter-only schedule.

Many believed, as did Chris Whittles and Chris Mead, that it was time to change tactics and start actually promoting feeding year-round. In reality, plenty of keen feeders were already doing so. To suggest such a practice to the general public, however, meant bringing the big bird and conservation groups on board. Understanding the complex and dynamic relationships between the birdseed suppliers and the BTO and the RSPB that followed would require a much more detailed treatment than we have time for here. To tell such a story would also require access to detailed and commercially sensitive information and insights well beyond the constraints of this modest exploration. What is well known—Whittles was characteristically candid about the arrangements—is that CJ Wildbird Foods was directly engaged with both BTO and RSPB for long periods (two decades in the case of the BTO), supplying a range of feeding products and collecting very substantial donations on behalf of both organizations. The Whittles-Mead relationship was central to this collaboration. Chris Whittles is especially proud of the part he played in establishing the BTO's Garden BirdWatch[20] program in 1995, though CJs had been involved virtually from the beginning of the company. Now an institution in Britain with thousands of paying participants, this project resulted from discussions between Chris Mead, Chris Whittles, and the BTO's then director Nigel Clark. CJ's sponsorship of around £20,000 each year ensured that the program had the financial stability necessary for the long haul. "This really got Garden BirdWatch off the ground," Whittles added.

The 1990s to early 2000s was the heyday for CJs and the British bird organizations, but it was not to last. "The arrangements with both the RSPB and BTO were always fraught because some people there always suspected my motives," Chris offered bluntly. "It's always about the money." During this period a variety of other players appeared on the scene, some specialist birdseed suppliers, others diversifying their product range to include a wide range of feeding-related products. "It's a crowded marketplace these days," Chris told me. "Far too many substandard offerings in my view, and price is the principal criteria! Where is the old-fashioned notion of quality?"

The Oddie Effect Rather than the proverbial golf-course deals, bird-food business negotiations were often developed at remote ringing sites around England, largely away from the public eye. The next event in this story was, however, very public indeed. In 2004, Bill Oddie, the extremely popular comedian and bird-watcher, hosted a three-part natural-history television series called *Britain Goes Wild.* The show was a spectacular success, becoming the BBC's most-watched natural-history program to that point. Having reviewed the parlous state of the British countryside for wildlife, Oddie passionately urged viewers to transform their own gardens into wildlife habitat. They could do this themselves simply by installing nest boxes and birdbaths, but especially by feeding birds. And not just in winter but, yes, indeed, throughout the year. It was all about providing a new bird-friendly habitat to replace what was now lost or degraded, and an alternative—and reliable—food supply. The television series was designed as a component of the broader BBC *Make Space for Nature* project, and an astonishing 83,000 people signed up in response to Oddie's appeal, with over 40,000 pledging to purchase a feeder! As a direct result, the RSPB and other outlets rapidly sold out of nest boxes and feeders. So significant was this single media event that a birdseed industry magazine article titled "How Summer Feeding Is Boosting Wild Food Sales" attributed the spectacular growth in demand directly to the "Oddie Effect."[21]

For UK feeders, therefore, the big change to all-year feeding seemed to have been successfully promoted by the bird and conservation organizations and bird-food companies by linking the conservation of small birds, especially the familiar species that came readily to garden feeders. The scientific data clearly demonstrated serious declines in populations of even familiar birds such as the iconic Blue Tits and even the (only recently disparaged) House Sparrow. The obvious response was to feed them. The assumption was that this would help. The evidence for *that* would hopefully come later, but now was the time to act.

Germany

The transition to year-round feeding was dramatically different in Germany. In North America, the bird-feeding industry has long been actively marketing a wealth of products to a public eager to simply attract birds anytime. In the UK, long-held traditional attitudes concerning when to feed have gradually given way to a widespread desire to assist birds throughout the year, a transformation that took place over several decades. In Germany, it's been a remarkably rapid *Völksbewegung*, a national people's movement.

Until relatively recently, the stance of the main German bird and conservation organizations, and that of the scientific consensus, was unequivocally one of winter feeding only.[22] This position was perhaps even more assertive in Germany than in other European countries. Indeed, after perhaps a century of organized and systematic bird feeding (already well under way prior to Baron von Berlepsch's appearance), by the middle of the twentieth century, the popularity of the practice appeared to have been waning significantly. Furthermore, outright opposition to all forms of bird feeding—but especially outside winter—was becoming more common among scientific and conservation commentators.[23] The general theme of these opponents appeared to be that there was little evidence that feeding wild birds had demonstrable benefits, that the risks—spreading disease, chicks being fed the wrong food—were serious issues, and that the only real answer to the alarming decline in European birds was to restore habitats. Furthermore, it was asserted, bird feeding really only assisted the rich bird food companies.

These were all familiar concerns and opinions, and they were being voiced by the most authoritative organizations in the country. The public who were aware of these discussions, especially the large number still actively engaged in bird feeding, were likely to be either confused or worried. Whether or not these attitudes actually influenced the behaviors of feeders, the debate was unlikely to encourage more people to feed. To find, therefore, that there had been a massive increase in the popularity of the practice, and with much of this being year-round feeding, all in less than a decade, is both astonishing and unexpected. This phenomenon seems even more unlikely when it appears to be attributable to a single factor: a white-bearded, retired biology professor in his 70s.

Professor Peter Berthold is a major figure in the international ornithological field, best known for his career-long research on bird migration.[24] His pioneering work on the changing movement patterns of Blackcaps, in particular, is justifiably famous (and of direct relevance to this discussion as outlined a little later). Although he officially retired from the Max Planck Institute for Ornithology in Radolfzell, southern Germany, in 2006, he maintains an active research role there in an emeritus position. He has been actively involved for many years in both German and international scientific bird arenas with positions on Deutsche Ornithologen-Gesellschaft and the International Ornithological Congress committees. In other words, Berthold has substantial scientific and academic stature. It would be equally fair to say, however, that he was not particularly familiar to the general public. Like most recently retired researchers, Peter Berthold probably dreamed of finally having time to complete that pile of unfinished manuscripts and working on the next edition of his classic tome *Bird Migration*. A quieter life, perhaps, with less pressure and more privacy.

But that was before *Vögel füttern, aber richtig*.[25]

In producing—and actively promoting—this book and its provocative stance, Professor Berthold has been propelled into the strange realm of scientist celebrity. Instead of fading into desired or inevitable obscurity, Berthold has become the "professor of feeding," astutely media savvy and unsubtly proclaiming his message. When a passionate and articulate elder scientific spokesman appears regularly in the media arguing in favor of continuous feeding, people are likely to take notice. When it's argued as fervently as it is in *Vögel füttern, aber richtig*, it is virtually impossible to ignore.

Although it's possible to regard this compact volume as just another popular "how to feed birds" book, *Vögel füttern, aber richtig* can be immediately distinguished from similar books by its scientific orientation and the audacity of its tone. There is a concerted attempt to provide a level of scientific evidence to support the arguments made, with numerous studies being cited and explained. That is expected, of course, in popular science writing, which this certainly is. What is less usual is the combative and, at times, aggressive, way this is done.

The context for Berthold and Mohr's proposal is the extent of environmental damage all too evident in Germany. According to these authors, it is undeniable that human activities have destroyed and degraded, polluted and poisoned the habitats required by birds, resulting in the familiar and tragic declines in so many species. Why? "The main reason is a decline in food availability!" In rhetoric and language reminiscent of that used by Baron von Berlepsch over a century before, Berthold and Mohr draw attention to the catastrophic impact of humanity on bird populations and advocate the strategic intervention of wholesale food provisioning. Of course we should also invest in restoring bird habitat, but they point out that "this is often harder to achieve than supplementary feeding."[26] Unlike von Berlepsch, however, who was stringently a winter-only feeder ("Birds *only* need feeding during and after certain changes in the weather, especially during blizzards and intense frost"),[27] Berthold and Mohr are evangelical year-round feeding advocates and would probably dismiss his concerns with the same disdain they reserve for other doubters. You are either with them or you are not.

The book is also unusually opinionated and intolerant, and indeed a bit triumphant. Reviewing the apparent change in public opinion brought about by the first edition of their book ("According to our daily mail, thousands of people have already shifted from winter feeding to all year feeding"), Berthold and Mohr observe that "since the start of the debate, opinion has shifted from misleading, emotional and idiotic opinions to soothing objectivity—to the great benefit of the birds!" Moreover, they attribute the residual opposition mainly to a benighted few who persist for "ideological reasons" and conclude (in translation):

> *Vögel füttern, aber richtig* not only provoked a shockwave towards feeding birds all year round, but also a wave that still gains momentum. To maintain the dynamic of the movement it would be beneficial to find out where in the country are the true bird friends and where are the hypothetical bird friends.[28]

It is probably too easy to focus on the writing style of *Vögel füttern, aber richtig*, and perhaps German readers do not register the tone as being rather abrasive and dismissive (at least in translation). More important, however, Berthold and Mohr make it their mission to promote feeding explicitly for conservation. The birds of Germany—and Europe generally—are in serious decline and strong action is required as soon as possible. Feeding, they declare, is a simple and effective way for ordinary people to address this crisis and an activity available to almost everyone. Hence the book and its forthright message: feed now and continuously.

It had become unavoidably clear that Peter Berthold was indeed one of the major personalities engaged in this part of our story, just as Chris Whittles had envisioned. If so, perhaps I needed to contact Professor Berthold, as I had contacted most of the other major players. After all, I had only been able to assess his attitudes indirectly, through the opaque window of translation. Maybe I had misinterpreted the tone and style. If I was to make these observations of such a central exhibit, I really did need to verify my opinions with the professor himself. I have to say, this prospect was somewhat daunting, given the professor's status and influence in the field. As a result, when I phoned late one night, I was happily surprised by the enthusiastic and positive response: Peter Berthold was only too willing to talk about bird feeding and especially his book.

I started the conversation with my most pertinent and possibly sensitive questions: *Why did you write the book, and why was the tone so, well, aggressive?*

Professor Berthold did not hesitate. "Oh, I was very angry! It may come through in the writing perhaps, but I had to address these old ideas about feeding. I had to be strong about this, so that ordinary people would listen." Berthold had been a conspicuous leader in the major German bird societies for years but felt he could no longer remain silent when these same organizations were so negative about feeding. "They simply did not understand that if enough people were to feed—in the right way—we could reverse some of the declines. Germany, and much of western Europe generally, has lost so much of the traditional food supplies—grain, weeds, fruits, berries, and also insects—yet we now can provide alternative sources of these foods. It is something we must do!"

I congratulated the professor on the remarkable sales of *Vögel füttern, aber richtig*. "Yes, astonishing figures! Currently over 150,000 copies have been sold; not so bad for a mere bird feeding book." This was staggering news; these were numbers usually associated with airport crime thrillers. "I was completely unprepared for this level of success," he went on. "And for the publicity that came. People—complete strangers—recognize me in the street, at the train station. I am often asked to speak about the importance of feeding on television shows and the like. And I always hold up the book and say: 'Read this please.' People say I am just interested in selling books. But it is always about the birds."

Initially, Peter's self-confessed anger had been directed to writing a simple journal article where he could outline his case for year-round feeding. "As I started to gather the evidence of the benefits of feeding, I began to realize we had more material than a small piece and it just grew into a book." Much of this material came from Peter's extensive contacts in Britain, including Chris Whittles. "Oh, I have been bird ringing with Whittles many times. I was very interested in using some of his products to attract birds to the nets. He even came to Germany in the early days to assist us in the Blackcap work. These birds loved his peanut-fat cakes, and our captures were much more successful. This was the 1990s and the first time I had really considered feeding in spring. It was an important time."

Purpose and Personalities

As I try to formulate some sort of conclusion to this exploration of the changes that have occurred when people feed wild birds, I am struck by two rather obvious themes. First, it depends on what people perceive as the primary "purpose" of their feeding. For perhaps the majority of places where bird feeding occurs around the world this purpose appears to be one of simply attracting birds for the pleasure of seeing them close up. It is more about enjoyment and appreciation rather than conservation or assistance. In other words, feeding may be regarded as a pastime, a leisure-time activity, a hobby. If so, feeding can occur anytime, regularly or occasionally, irrespective of time of the year. For these feeders, this earnest discussion about winter feeding versus year-round is probably redundant—and probably irrelevant. This seems to be the case for many feeders, but especially in places where climatic or environmental conditions are mild; while the birds readily take the food being offered, they don't really need to. In these circumstances, there is typically little discussion about when feeding should happen.

On the other hand, people may be feeding birds with the intended purpose of helping. This is particularly evident in places with severe climatic conditions, where birds are perceived to be suffering, especially during harsh and prolonged winters. As we have seen, feeding may also be associated with assisting birds by replacing diminished food supplies and increasing their likelihood of survival. In these circumstances, the reasons for feeding would appear to be much more significant than a mere pastime. Feeding as helping is, therefore, much more likely to lead to discussion about possible benefits and risks—because this is not just about human enjoyment, it's all about the birds themselves.

Now, this dichotomy is clearly far too simplistic (and I will be outlining just how complex our motivations for feeding can be in a later chapter). It may, however, assist us in attempting to discern the different attitudes evident in North America compared to those in Europe, at least in recent times. While feeding as attracting and helping obviously occurs in both regions, the differing conservation status of birds using feeders appears to have conferred a contrasting perspective on the practice. We could say that North Americans feed birds to *see* them, while Europeans feed birds to *save* them. This may offer some explanation as to the relative nonissue of when to feed in the United States and Canada and the strength of the debates around why and why not seen in the UK and the rest of Europe.

I was also struck by the powerful influence of a relatively small number of people on the change in the seasonal timing of feeding that has occurred in Europe. From all the material and personal impressions I have been able to glean, it seemed that the big change was far more influenced by these personalities than by the scientific evidence. There had been references to certain, sometimes unspecified, "studies," but remarkably little actual data. Part of this will relate to the lack of research undertaken prior to the crucial 1980s–1990s period. Some important research had been conducted (including the "forgotten" Thompson studies), and it provided significant insights into crucial factors, such as whether parent birds used feeders to gather food for nestlings and how feeding could influence survival. But providing convincing scientific evidence seemed less of a factor in bringing about this profound change in the practice of feeding than the exhortations and forceful statements of a few well-known spokesmen.

That the personal opinion of effective and respected authorities has been more influential than straightforward logic and argument is not really that surprising. How the message is sold has been at least as important as the message itself. For this story, however, it is time to have a closer and more critical look at the evidence associated with wild bird feeding. A lot of claims and definitive statements have been made along this journey so far. Just what do we know with certainty about the effect of all those feeders?

4

The Feeder Effect

What All That Food Can Do

November may not be the ideal time to visit the northern part of New York State, but the spectacular late-fall colors do compensate a little for the biting cold. Massive snowfalls across the Great Lakes region had threatened to stall my travel plans, so I was relieved to see the rolling landscapes around the Finger Lakes ablaze in vivid oranges, yellows, and reds as the small plane turned to descend. There were many reasons I was looking forward to getting back to the small rural city of Ithaca: seeing several long-term friends, wandering through the spacious campus of Cornell University, trying varieties of sweet red apples I had never heard of, enjoying the hoppy flavors of the locally brewed beer. My family and I had spent a delightful and productive sabbatical based at the university almost twenty years ago. Significantly, that period involved my first experiences as a bird feeder, American style.

In Ithaca, we lived in a large house with a sprawling garden, complete with five separate feeders. The owners, themselves academics away on sabbatical leave, left written instructions on every aspect of home and garden maintenance, but those associated with the feeders were particular detailed. The garage contained various drums of black sunflowers and several seed mixes, along with all manner of funnels, ladles, and containers for carrying the food to the feeders. As I surveyed the equipment and read the itemized instructions, I began to appreciate that this was something of a big deal. As someone with no experience of feeding at all at the time, I admit that it took me a while to "get it." As I began to engage in the process of filling the feeders and watching the results, my reactions developed roughly in the following order: indifference (*Why would anyone be this interested in feeding birds?*); responsibility (*I had probably better follow these instructions carefully*); tentative fascination (*Actually, this is quite interesting*); and finally, sincere commitment (*I'll just top up those three back feeders again. It's getting pretty tough out there!*). At the time, however, I probably thought of it as simply a temporary "vacation experience," soon to be little more than a fond memory. I now realize that it was an experience with lifelong effects.

Even if we start feeding for the simplest reasons ("I just like seeing them close by"), the possibility that providing food may be influencing the birds in various ways is soon brought to our attention. Whether this is from fellow feeders, neighbors, the media, or official advice from bird and conservation groups, there are plenty of claims about the ways that feeding may be affecting the birds. These are often positive such as improving their welfare during times of scarcity, enhancing survival, and supporting populations of declining species. But there are also those negative claims: the birds may become dependent on the food we provide; concentrating birds at feeders can spread disease; and feeding is leading some species to stop migrating. These, and lots of other claims, are potentially significant and are certainly worthy of our attention. Such issues are clearly important for a book such as this. While some have already been mentioned, understanding how much we really know about the "feeder effect" is central to this journey.

Which brings us to the unsettling issue of evidence. This is unavoidable because we are attempting to take a critical and scientific approach to assessing some of the claims and statements associated with feeding. A scientific approach means finding and evaluating the relevant research and synthesizing it into a form that is both faithful to the science yet accessible to the reader. There are many studies on a wide variety of species that look at how the addition of food influences their ecology and behavior. These experimental studies, known as supplementary feeding research, is extensive, and the findings are discussed in detail in Chapter 5. Many of these experiments provide important findings of relevance to us. However, almost all have involved strictly controlled research focusing on single species conducted in natural environments, well away from people. This is important work but is very different to the delightful chaos of a typical suburban garden where feeding occurs, with all the different feeders, foods, species, practices, and variability. No, attempting to understand the broader influence of wild bird feeding as it occurs across large areas is going to require a very different approach.

The reality is that while wild bird feeding may be a massive enterprise with a global reach, the focus and interest of most individual feeders is on their own private garden. All those issues and worries may be of potential interest, but unless they affect us directly, they may seem a bit too vague or remote. On the other hand, most feeders are pleased to share their experiences or queries (such as that strange behavior I saw yesterday, or, Did you know that [species *x*] ate [food *y*]? Or, Is it unusual to see [species *z*] here at this time of the year?). I am delighted to share my personal observations, especially if I know it helps understand something of the big picture. But I guess I would need some convincing as to whether it was worth my while.

Convincing an eager army of willing feeders to turn their personal observations into precious data has been the astonishing achievement of a program called Project FeederWatch.[1] Since its modest beginnings in the 1970s, the information gathered from ordinary people has transformed what is known about many of the birds of North America. Can the project also tell us about the effect of feeders across this vast landscape? The people who run Project FeederWatch are based in Ithaca, and I am here to learn more.

A Visit to the Lab

Just outside Ithaca, surrounded by woods and lakes, is the internationally renowned Cornell Laboratory of Ornithology, typically referred to with warm informality as the "Lab of O." Through its close links with Cornell

University, the Lab's academic credentials are undeniable, but arguably its most significant achievement has been in attracting huge numbers of regular people, not for degrees, Ivy League reputations, or even to mingle with some of the greatest minds on the planet, but simply through a shared appreciation of birds. The variety of approaches taken by the Lab to achieve this has been astonishing and often technological—bird ID apps, easy-to-use sound-recording software, instantaneous logging of bird locations, for example. But it has been the sincere and effective engagement with vast number of people that really stands out.

The Lab of Ornithology did not invent the concept of "citizen science," but they are leaders in putting it into purposeful, effective action.[2] Utilizing the time, enthusiasm, and availability of ordinary people in order to observe, record, and participate in any number of science projects has been an almost revolutionary movement in recent times. Of course, the challenges of being able to transform the initial willingness of citizens into a meaningful and ongoing partnership with researchers are formidable. Inadequate training can all too easily result in useless data, while too much control—or too little—can discourage and disappoint the participants.[3] It's a fine line, as the many failed citizen science projects around the world can attest. The Lab must be doing something right: each year enormous numbers of people count, record, watch, and then send in a mountain of priceless data. I'm here to find out why.

The walk from the parking lot to the front door of the imposing Imogene Powers Johnson Center for Birds and Biodiversity, home to the Lab, is not that far, but at –17 °F and in a bitter wind, even I forgo the chance to scan the nearby Sapsucker Woods Pond for sheltering ducks. Once inside, however, I realize that the entire western side of the two-story building consists of glass walls, looking directly out over the pond. Well, what did I expect? This workplace is literally full of birders. After I give my name to the attendant at the reception desk, she hands me a pair of excellent binoculars and, waving in the general direction of those massive windows, says, "Check out the waterbirds while you wait." I do as instructed: Hooded Mergansers, Wood Ducks, and Mallards are out on the choppy waters and even a solitary, silent Great Blue Heron stands statue-like nearby.

A sudden flutter of much smaller birds draws my attention to the bare shrubs just outside the window. They are swarming to one of the numerous of feeders of various designs positioned within view: hanging tubes of different widths, covered platforms, and suet baskets. Restless groups of little birds land on the feeders, jostle for a spot, then move on to the next. I think I see Black-capped Chickadees, American Tree Sparrows, Red-breasted Nuthatches, and an American Goldfinch, but they are flitting about so relentlessly it's hard to be sure. They certainly look hungry, and it's obvious that competition is fierce out there in the freezing wind. I can't help but wonder: What would they do without the feeders?

My birding is interrupted by the arrival of two people I had wanted to meet for a long time: David Bonter, who had coordinated Project FeederWatch for over ten years, and Emma Greig, who took over from David a few years ago. Together they should be able to provide long-term perspectives as well as current realities on one of the most successful citizen science projects in the world. Before we turn to leave, David scans the sky, and Anna the closest feeder. "Lots of snow tomorrow," and "That feeder needs refilling," they observe respectively.

Project FeederWatch actually started in Canada as the Ontario Bird Feeder Survey.[4] The Long Point Bird Observatory (located on an extended finger of land jutting far out into Lake Erie) recruited what seemed like a lot—333 to be exact—of "kitchen-window researchers" for the first trial during the winter of 1976–1977. The aims were modest: to get people to record which species were coming to their private feeders, how many, and which food they liked best. There was plenty of skepticism, especially from "real" scientists, who queried the skills of these untrained observers and therefore the reliability of the information they gathered. These are valid criticisms and were taken seriously by the organizers of the program, leading to continuous refinement of the instructions and feedback to participants. But the numbers of people enthusiastically involved continued to grow and spread, especially in the US Northeast. At around the ten-year mark, the Canadian-based survey joined with the Cornell Lab to provide truly continent-wide coverage. Renamed Project FeederWatch and coordinated by Erica ("Riccie") Dunn and from then on headquartered at the Lab, the new combined program was launched in the winter of 1987–1988 with over 4000 participants.[5] Currently—almost forty years since it began—the project has over 15,000 members, all contributing high-quality information, on a scale utterly impossible by any other means.[6]

"Yes, those numbers are impressive, of course," explains David Bonter, "but the really telling statistic is 'seventy,' as in 70%. That's the retention rate of people in Project FeederWatch, meaning that seven out of every ten participants continue to stay involved the following year. The average for most citizen science projects is less than about 10%. Everyone is always enthusiastic when they start something, but it is really hard to keep them interested year after year. It seems to be different with FeederWatch; folks just love their birds."

I suspect that there is a bit more to it than that. I ask Emma Greig, the current coordinator, why she thinks people are so loyal. "I think that they join up because they are already interested in birds and they stay involved because they feel that they are part of something big and important, that even though they may not be scientists themselves, they are contributing to a major science project. But maintaining that perception requires constant feedback; it's a lot of hard work!" This involves, among many other things, keeping a superbly interactive website up-to-date and looking continually fresh. New stories are needed constantly, the incoming data needs to checked and rechecked, and decisions made about the veracity of reports. ("A mockingbird in Washington State?" Emma says. "Really?") The Lab makes its mountains of data available to scientists and the public, for research and curiosity, so the data have to be thoroughly reliable. These kinds of quality control and checks are a normal part of any scientific research, but it's often the "citizen" part of citizen science that takes the time. For instance, a big part of Emma's day-to-day work is hands-on, old-fashioned phone conversations with FeederWatch members about mundane problems and personal requests. Plenty of folks still like to chat about the "lovely little bird that just showed up" or to renew their membership over the phone. If Emma occasionally feels overwhelmed, she doesn't let it show. Which is just as well because, I am shocked to learn, this continent-wide, mass-participation, internationally renowned program is managed almost entirely by Emma and a couple of part-time assistants.

As a result of successfully recruiting, training, and retaining an enthusiastic multitude of members for several decades, Project FeederWatch has received a vast amount of data. With diligent observers located throughout—though not evenly—the entire North American continent sending in monthly records of over a hundred species that visit feeders (from November to April), extremely detailed information is now available over a massive landscape. This provides a clear and reliable picture of the dynamic nature of changes in movements of feeder birds in both space and time.

Actually, given the scale of the geography and time span, the "picture" is more like a video that represents changes in time as well as space.

David Bonter invites me to sit beside him at his desk so I can view his large computer screen. He demonstrates the ease with which anyone can interrogate the Project FeederWatch database to visualize changes in the distribution of any particular species over the whole of North America.[7] "Let's pick a winter favorite, the Northern Cardinal," suggests David, not knowing that a glorious photograph of this bird—vividly scarlet against a snowy background—is an incongruous feature of my office wall back in Brisbane. In the "Map Room" section of the Project FeederWatch website, David selects the annual February totals for cardinals, and within seconds a map of North America pulsates as millions of accumulated reports are displayed as colored dots. The map background is blotchy with grey circles indicating the location of project participants, while places where cardinals had been recorded glow yellow. As expected, the sweep of yellow conforms to the general distribution map for the species, as shown in any bird guide: cardinals are found at feeders throughout the eastern half of the United States (apart from a strangely separate population in Arizona) and just into the southeastern provinces of Canada. The maps of bird distribution in the guidebooks suggest that the area occupied by a species is somehow fixed. Switching to a different display showing changes year by year, however, shows clearly that this idea is far too simplistic.

"This is a particularly interesting area to watch for cardinals," says David, pointing to the eastern US-Canada border region. As the animation advanced gradually from 1989 to the present, the spread of yellow dots, slowly though unsteadily, pulsates northward until the data stop at last year's records; yellow dots are now spread all over Nova Scotia and the Maritimes as well as central Quebec and Ontario. What this demonstrates—explicitly—is that Northern Cardinals are now seen regularly in February much farther north than just a couple decades earlier. "Cardinals are iconic and everyone wants to see them in their gardens, so their movement north in winter—or at least their willingness to stay put during the depths of winter—has been of particular interest to Project FeederWatch participants," explains David. The same thing is happening with less charismatic species too: Tufted Titmice, Carolina Wrens, Red-bellied Woodpeckers, and others have all shown a similar pattern of being observed regularly farther north than their traditional distributions.

The obvious question is "why?" We need to be careful: the most obvious answers are not necessarily the best explanations. As tempting as it might be to claim that these spatial patterns are all about feeders—or, indeed, to maintain that it has nothing at all to do with feeders—the reality is likely to be a lot more complicated. Certainly, many observers and feeders have noted the northward movements of the cardinal over recent years, but a major range expansion of this species has been under way since the late 1800s. The first Arizona Cardinals were recorded in the 1870s. The birds apparently first reached Toronto during the 1930s and spread west into Minnesota around the same time. These significant movements were occurring long before bird feeding was a common practice, though there is equally no doubt that feeding has played an important role. A recent study of cardinals in Ohio found them to prefer urban areas to other habitats in colder periods and concluded that this was probably due to general characteristics of cities such as dense undergrowth and warmer microclimates.[8] The availability of food was found to be particularly important in winter, but, perhaps surprisingly, fruit-bearing trees and shrubs had a much greater influence on the presence of the birds than the number of bird feeders. In other words, feeders are just one of the features that cardinals weigh up when they are assessing a location. Obviously food supply is a fundamental factor, but it is not all about the stuff humans provide. Naturally occurring berries, seeds, buds, and fruit—including those found in gardens—are always going to be crucial; the significance of feeders will probably depend on the severity of the conditions.

In cold conditions places with concentrated human activities are also warmer than the woods and countryside that surround them. Many species—including the cardinals of Ohio—do not have to read the scientific literature to appreciate the heat island effect of towns and cities and move in to take advantage of the milder temperatures in winter. This is also an effect occurring on a global scale. Everywhere, as the climate warms, thousands of species are steadily shifting their distributions poleward, as the historical climatic barriers associated with the cold move farther north (or south for the Southern Hemisphere, or up mountains). This is not the place for further discussion of this profound global phenomenon, though it is important to note that huge numbers of birds, mammals, amphibians, reptiles, fish, and insects are on the move. For our purposes, climate-induced changes in bird distributions add yet another possible influence to the mix.

David Bonter draws my attention back to the screen in front of us: a new animation is showing the Project FeederWatcher records of the Carolina Wren, a small, seemingly delicate species, gradually moving above the Canadian border since the turn of the century. "We always knew that this species sometimes ventured north a little in mild winters, but this data shows that some of them are now living permanently in the Montreal area—and part even further north—in February!" That's a significant shift in their range in little over a decade; something significant is behind movements like these. The Carolina Wren is a largely insectivorous species, a bird far less interested in the seeds of typical feeders, though they are partial to suet and fat balls. For Carolina Wrens, the influence of feeders may be even lower than for other species, but we can't be certain yet. "To some extent, it's the same with all the birds moving north. Is it climate, garden resources, natural food supply, warmer city environments? All of these? Can we actually separate them in any meaningful way?" David's candid questions suggest an unsettling mix of scientific curiosity and practical difficulty; these are really important queries but so hard to investigate.

David Bonter pushes back his chair and continues on this theme of thwarted ambition. "My biggest professional frustration with Project FeederWatch is in its extraordinary success in collecting enormous amounts of data but the lack of time and resources we have to do the necessary research that could answer so many really important questions. I have so many half-written manuscripts but no time to finish them! But it is just as much my fault too: I'm always trying new ways to investigate birds on feeders." Suddenly, David is describing a recent study, his frustration forgotten in his enthusiasm to describe what he has recently discovered.

As anyone with a feeder knows well, the number of birds making use of the food provided can change dramatically throughout the year. Sometimes there are clouds of them bickering for a perch or space on the platform, while at other times there is hardly a visit all day. Understanding the reasons behind this variation is one of the fundamental dimensions of bird feeding research, but the challenges are formidable. One of the main difficulties is knowing what individual birds are getting up to. Do certain birds visit continuously, others only occasionally? Or is there a constant flow of new birds, stopping briefly before moving on? To be able to address these sorts of questions, we need to be able to recognize individuals. Elsewhere (see Chapter 5) I describe the work of Margaret Brittingham, who transformed our understanding of the behavior of Black-capped Chickadees at feeders in winter. Her insights were only possible because she was able to distinguish between individuals, based on the color-banding (or "ringing") of hundreds of birds. This is standard practice among ornithologists today, but it only works if it is possible to detect and "read" the tiny, plastic colored stripes on a constantly moving bird. That can be hard work, limited by human sight and the availability of observers. But what is happening when the researchers are not around?

A new approach to this problem is to have the feeding apparatus itself automatically record the visits of individuals birds. Using technology similar to the PIT tags most of our pets now carry, minute, individualized electronic tags can be attached to the leg bands of the birds. When these tags are detected by a receiver set up at a feeder, the identity of each bird's individual radio frequency is logged, providing a precise record of each visit of each tagged bird. This technique, known as radio frequency identification (RFID) is already providing extraordinary new information on the foraging behavior of birds in relation to feeders. "Let me illustrate what I mean," said David, pulling out a writing pad and pencil. "We tagged a big sample of chickadees in a woods near houses and fixed the RFID receivers to feeders in back yards nearby. Our main interest was in the use made of feeders over the whole year, a timescale that would be very difficult simply using human observers. The technology is so robust and the tags so reliable you can set up the system and it hums away for months, diligently logging every bird."

On the pad David sketches a graph showing the visits per day for the first year of the RFID study. The rate of visits was typically moderate from January and dipped even lower during late summer. But in August and September, daily visits suddenly shot up dramatically, with David's drawing showing a huge and abrupt spike, before falling just as rapidly to "normal" levels by October. "This spike coincided perfectly with our prediction of the prewinter seed-caching period. These tiny birds are preparing for the cold that is just around the corner by storing seed away. When I saw these data I remember thinking: 'We are so good!' This is just the sort of evidence we needed to demonstrate the significance of providing supplementary food for chickadees. But I should have known it is never that simple. In the second year, it was as though all of the feeders (and all those expensive detection gadgets) just didn't exist; instead of that spike in visits we saw in the previous fall, this year numbers stayed way down. What were they doing instead? Feasting on the massive harvest of ash tree seeds that occurred that year! It was a sobering lesson in assessing our influence on the lives of these birds. Again, thankfully, the birds were sensibly assessing and utilizing all the available foraging resources, and this year, it was all about the natural stuff." Sometimes, it seems, feeders are crucial; at other times, trivial.

Sometimes, on the other hand, feeders really appear to be essential for the very existence of a species in certain locations. As another example of the changing distribution of a bird species revealed by Project FeederWatch data, David selects the February records of a West Coast favorite, the exquisite Anna's Hummingbird. Although their stronghold is in California (where 75% of project participants have hummers visiting their gardens), their distribution has been extending ever northward. The animation shows patchy sightings in Oregon and Washington during the 1990s, but this fills out in the 2000s, and by 2014 Anna's have a solid presence throughout the western coast and well into southern British Columbia. "This is certainly one of the strongest movements from the FeederWatch records," says David. But the reasons underlying the phenomenon seem a little clearer for these hummers: the widespread planting of winter-blooming ornamentals and the sheer abundance of hummingbird feeders. A little consideration of the map and the continuous yellow dots spreading into lower eastern Canada—in the depths of winter—brings up the incongruous image of exotic hummingbirds flitting around a snowbound landscape. Friends from Seattle recently sent photos of Anna's Hummingbirds amid large icicles, sipping eagerly from a specially heated sugar-water feeder. This was such an unlikely, almost unnatural, scene, probably only possible through direct and sustained human intervention. The admittedly marvelous sight of hummingbirds in the bitter cold also generates its own questions: Would these birds even be there without the feeders? Should they be? These birds are famously territorial, so what happens when a feeder freezes or is neglected? Do the feeders actually improve their chances of survival over the winter? And given that hummingbirds are adored everywhere, and that people will never stop wanting to see them—even (especially?) in colder parts—does it really matter?

"There are just so many important questions to ask," says Emma Greig, as we sit in an observation room overlooking Sapsucker Woods Pond. The pond, which was hosting dozens of waterbirds only a few days earlier, is now fully iced over. A few hardy ducks remain, hunkered down in the drifts. It is bitter out there, and the numerous feeders in sight are busy: chickadees, Tree Sparrows, House Finches, and nuthatches apparently amicably sharing the bounty. "Look at these poor birds. We sometimes feel sorry for them because we are safe and warm inside while they are stuck out there. But they are used to these conditions. They have evolved to cope with all sorts of weather or they wouldn't be here." This simple observation raises one of the really big questions that continuously circulate around the effect of feeders, and an issue that Emma is constantly asked about: Does feeding lead birds to abandon migration?

"Oh yes, that's a big one!" Emma sighs and glances out at the feeders. "Certainly, there are examples of migratory birds staying put through the winter—Canada Geese are probably the best-known example—and feeding has been suggested as one of the reasons. But in almost all of these cases, it seems to be mainly some individuals rather than a whole species or population that are remaining behind, or perhaps groups of birds in specific circumstances. I can't think of any species or even population where migration has just been abandoned entirely." Emma pauses to reflect for a moment, possibly recalling her time spent in Australia where she completed her doctoral studies on tiny fairy-wrens. Like the vast majority of Australia's avifauna, these birds are decidedly nonmigratory; most are either permanently sedentary, living their entire lives in the same place, or are nomadic. The epic movements of huge numbers of birds north and south, so familiar and predictable throughout the Northern Hemisphere, are just not a prominent phenomenon in the southern parts of the globe, although the long-distance movements of waders and shorebirds are notable exceptions. My interpretation of Emma's expression suggests that she is considering a quick tutorial on bird migration for the visiting antipodean: Bird Migration 101.

The extraordinary migratory journeys made by many birds are ultimately controlled by the internal hormonal and physiological cycles of the birds. These inner cycles, which function daily as well as seasonally, are fairly strictly linked to grand external cycles such as the slight but utterly rigid variation in day length. But birds are not robots; no year or season is exactly the same, and they must be able to alter decisions about when to act (start building up reserves for the flight, or gather in groups, for example) if the conditions are not suitable. There are many such potential influences that may trigger these migratory behaviors in birds—temperature, changes in vegetation, food supplies—all of which can vary from day to day and year to year. If the birds were to respond primarily on these fairly volatile signals, they could fly off too early or too late, both with big risks. "After all," explains Emma, "the reality is that they need to time their travels correctly to the breeding areas or it may be too cold, too early, or too late when they arrive. Similarly, once breeding is over, they need to move at a suitable time to maximize the survival chances of their young but avoiding the coming cold, storms, and general lack of food. To get the timing right, most songbirds seem to have evolved over the eons to respond to the small but crucial changes in day length as they prepare and eventually leave."

But this otherwise sound, reliable system seems to be coming unstuck. Across the globe, birds of all types, from tiny warblers to large shorebirds, are changing their age-old traditions, once regarded as fixed and immutable, and are arriving increasingly earlier in spring as well as delaying their autumn departure dates. This general pattern is most clearly being attributed to the warmer weather of the past few decades; yes, climate change is also changing ancient migration patterns. This seems to suggest that the apparently fundamental day-length trigger, which does not alter, can be overridden by the influence of temperature. "Just how far this might go is really unknown at the moment " says Emma. "But it is certainly going to affect the pattern of migratory birds using feeders. What used to be traditional arrival times for all sorts of species are already very different from a decade ago. It's a big point of discussion."

Emma pauses, perhaps regretting the decision to summarize several centuries of international research so succinctly. I, however, am grateful for such a concise explanation. "Given that grand scenario," I offer, "feeders do seem to be less significant than other influences that migrating birds are assessing." Emma agrees. "Foraging resources are always going to be of critical importance, but it you are a small and vulnerable species, winter is likely to be just too tough. Better to head south, even if there is plenty of food. It would not be a sensible idea to become overly reliant of any particular food type, including feeders." She pauses, looks thoughtful and adds quietly, "Of course, there are plenty of apparently small and vulnerable species which don't move and are even moving further and further north. Oh, it seems so complicated, but that is what you get with functioning ecosystems. To survive birds must be able to adapt to changing circumstances. They have been migrating back and forth for millennia. People and their feeders are really very recent additions to the landscape, so it's probably too early to know whether all that additional food is going to have large-scale evolutionary effects."

Our conversation returns to the difficulties in trying to isolate the possible effect of feeders from all the other influences, human and natural, that play a part in the way that birds interact with feeders. From a scientific research perspective, it is virtually impossible to control all these factors with any meaningful precision, especially in complex places like towns and cities. This is one of the reasons that much of the research on bird feeding has been conducted away from people and where true experimental conditions can be imposed. This simplifying of the variables is what defines the supplementary feeding experiments that will be discussed in the next chapter. Standard experimental design requires a "control" where everything is the same as the "treatment" but without whatever that treatment might be. The big problem for those of us trying to discern the effect of feeders is that control sites are difficult to find. As David Bonter exclaims, in places like the Northeast and along the west of the United States, "feeders are everywhere!" Elsewhere, this is certainly not the case. Assuming that the distribution of Project FeederWatch contributors reflects that of bird feeders generally, the map of participants across North America (easily found on the PFW website)[9] shows that the density of feeders tends to mirror the broad density of people across this vast landscape. This variation in the number of feeders per area actually provides an opportunity for investigating the feeder effect on a grand scale.

Tracking a Disease

One of the most significant contributions of Project FeederWatch to understanding major ecological and conservation phenomena involves the now famous case of tracking a serious eye disease that affected House Finches. Indeed, this example of large-scale monitoring combined with detailed scientific analyses is probably the most important example of well-organized citizen science in action.[10] The details of this extraordinary collaboration of feeders and disease scientists over several decades will be covered in Chapter 6. For the moment, however, this phenomenon provides a valuable opportunity for assessing the possible influence of feeders on a species at a large geographic scale.

Formerly common in the drier areas of the western United States, the House Finch now found throughout the east apparently originated from birds released in New York in the 1940s. Starting in the 1960s, the species began to spread steadily westward with the eastern-introduced birds eventually overlapping with the original population on the western plains in the 1990s. This rapidly expanding range was recorded with remarkable precision as diligent Project FeederWatch observers noted the arrival of what was a new species. The birds would have been hard to miss: they are highly social, foraging in big flocks and often dominating the traffic at feeders.

Then in the winter of 1994, people began to notice House Finches with swollen, encrusted eyes. With remarkable foresight and resolve, the Cornell Lab of Ornithology launched the House Finch Disease Survey almost immediately, utilizing the vast capacity of the Project FeederWatch membership.[11] As a result, the progress of the disease through the eastern population was mapped with unique precision and detail. It soon became evident that the rapid response was justified: this bacterial infection was highly contagious and lethal. The number of House Finches in the worst affected areas declined massively during the late 1990s. Reports of diseased birds began to level out by the turn of the century. Recovery, however, has been decidedly patchy throughout the eastern states, with the abundances of House Finches in some places now back to pre-outbreak levels, while in others the species remains rare.

The role of feeders in this catastrophe and the aftermath has long been debated.[12] House Finches' tendency to feed in flocks and aggregate in high densities at feeders obviously increases the likelihood of infected birds spreading the bacterium to others. On the other hand, because the disease impairs the bird's eyesight, the predictable location of a feeder and the reliability of finding food there compared to foraging in natural areas could potentially allow populations to recover quickly.[13] Given the rapid impact of the disease and the dynamic nature of the subsequent response over a large area, understanding the long-term relationship between feeders and abundance of House Finches would be extremely valuable. To achieve such a level of knowledge would require data on both feeder density and House Finch numbers from a huge geographical area as well as over a time span covering the outbreak.

Just such information is available, if you know where to look and are able to execute the appropriate analyses. Jason Fischer and James Miller realized that a measure of the density of people feeding birds was available from the numerous National Surveys of Fishing, Hunting and Wildlife Recreation undertaken by the US Census Bureau (discussed in Chapter 2).[14] For the abundance of House Finches they used data from the annual Christmas Bird Count,[15] a collection of counts of birds in a 24-kilometer radius during one full day in winter. Many thousands of people from everywhere in the United States participate, providing a high-quality snapshot of the presence of birds in winter for the whole country. Using the four national surveys that covered the period of the outbreak—1991, 1996, 2001, and 2006—the researchers were able to explore several key questions. Before the disease, did the density of feeders predict the number of House Finches? After the disease struck, did the habit of feeding in flocks—the likely cause of the increase in infections—result in lower numbers (as has often been claimed)? Or did the species recover more quickly in places where feeder numbers were higher?

Using data from twenty-two contiguous states of the eastern United States, Fischer and Miller obtained some of the strongest results yet of the feeder effect. Prior to the disease, the density of feeders was strongly related to the abundance of House Finches, suggesting that this species was clearly benefiting from the provision of all that seed. Perhaps House Finch numbers over large areas were linked to the number of feeders, more than they were for other seed-eating species. Interestingly, even after the disease hit and numbers declined significantly, the relationship of more feeders, more finches persisted. In other words, there was no clear evidence that the disease was more prevalent and had resulted in greater impact where there were higher densities of feeders. Furthermore, the growth of the finches' population following the fading of the disease and the restored relationship between feeder density and bird numbers seemed to demonstrate that feeders were providing a positive influence.

Although this study is limited to a single species in a rather atypical situation, it is important in providing clear support for the proposition that feeding is capable of leading to changes at large-scale population levels. While such a conclusion will be either welcome news for some, and possibly regarded as self-evident by others, Fischer and Miller have chosen to highlight some of the more sobering consequences of their findings.[16] Noting that the ramifications of these results were "profound," the researchers point out that while the House Finch appears to be recovering, the species is known to be capable of infecting other feeder species such as Purple Finches, House Sparrows, and American Goldfinches. This is a thoughtful corrective to our tendency to concentrate primarily on what we want to see. As someone mentioned to me while I was in Ithaca, "The Lab is famous for mapping the finch disease as it spread from feeder to feeder. But would it have spread at all if there were no feeders in the first place?" It's a moot point; feeders are here to stay. Perhaps a better question might be: "What can we learn from all those people watching their feeders?"

British Blackcaps

There is only one other program of a similar scale and significance to the North American Project FeederWatch and that is the British Trust for Ornithology's Garden BirdWatch.[17] While many other similar citizen-based garden bird surveys have been developed around the world, these two are the clear leaders. And as both are focused firmly on gardens, have huge numbers of participants (who each pay for the privilege) regularly submitting vast amounts of reliable data, and cover countries of particular interest to the feeding story, they are programs of central importance to the objectives of this book.

When I meet with Kate Risely, the coordinator of Garden BirdWatch, and Kate Plummer, a researcher delving into the riches of the monumental dataset at the BTO headquarters in Norfolk (described in Chapter 3), I asked (among many other things) the same naive question I had posed to David Bonter and Emma Greig in Ithaca: "Is it possible to show whether feeding is influencing birds?" The response was largely as I had come to expect: "Well, to some extent, yes," said Kate Risely. "But while there is a lot of detailed information on bird numbers, movements, and timing, it is still very difficult to separate out the influences of climate change, habitat loss, farming practices, and lots of other potential factors." It was a familiar story, and one that usually led into a discussion about the importance of carefully controlled feeding experiments.

But then there was a pause in the conversation as Kate Risely and Kate Plummer exchanged a possibly conspiratorial glance. "Although. . . . ," Kate Risely hesitantly began, "Kate has just come up with some very new results that you might find fairly interesting," she said. "Do you know the Blackcap story?"

The Blackcap is a much celebrated little bird, an otherwise typical warbler but with an unusually complicated pattern of migratory movements. During the breeding season, pairs of this neat, proper-looking warbler (the sexes conveniently differentiated by the smart, tight-fitting black skullcap in males, orangey-red in females) are found throughout the shrubby woodlands and riverine habitats of central Europe and the lowlands of the British Isles. As the weather cools, the birds that have bred in the UK and Ireland head south, to overwinter in the milder climes of Spain, Portugal, and northern Africa. While most undertake this fairly typical migration, it has been known for some time (as far back as the 1800s) that some Blackcaps could still be found in the British countryside, despite the wintry conditions. These birds wintering in Britain survived mainly on natural foods—berries and fruits—but gravitated to gardens to access feeders as the cold weather advanced. This late-winter move into gardens allowed keen feeder watchers—especially Garden BirdWatch participants—to record these visits over extended periods of time. Everyone assumed, pragmatically, that these Blackcaps were birds who had somehow "missed the boat" to the south and were therefore stranded for the winter. It was hard to escape the conclusion that they were somewhat lower-quality representatives of the species.

When the number of these overwintering birds recorded in Britain began to increase steadily, from just a few in the 1960s to becoming a regular garden visitor in the decades that followed, other explanations for this phenomenon began to be sought. The German migration authority Professor Peter Berthold (more latterly of all-year feeding fame, as discussed previously) took a particular interest and began ringing Blackcaps at his study site near Lake Konstanz in the late 1980s. The first astonishing discovery was that, rather than these wintering Blackcaps being stranded individuals somehow left behind, the majority were actually birds who had flown northwest from their breeding areas in central Europe.[18] This was a dramatically different direction from that taken by other European Black-caps, which was solidly and sensibly southwest toward the far warmer Iberian Peninsula. Why would some individuals of the same species be heading in such different directions? Traditionally, in this region moving sideways in winter rather than toward the warmth would be disastrous; they would be prolonging their exposure to cold and reducing the chances of survival. Nonetheless, the records from England showed numbers to be increasing, especially since the 1990s. Something rather strange was clearly going on.

With increasing numbers of ringed birds returning to their normal breeding grounds in southern Germany, Peter Berthold was able to study their reproductive activities in detail. He soon found that the Blackcaps that had overwintered in Britain—ostensibly the individuals taking the "wrong route"—were arriving back well ahead of the birds coming from the traditional south, often a full two weeks earlier. This was a huge advantage to these individuals, allowing them to find a mate, breed earlier, lay larger clutches, and produce significantly more offspring that those migrating along the traditional route.[19] Recent genetic studies have confirmed that the overall European Blackcap population is progressively separating into two groups: those wintering in the UK and those heading south, with males and females from either group preferring to mate with partners with the same migratory inclinations.[20] This is a dynamic picture of evolution occurring in real time before our eyes.

Fascinating though this unfolding story is, a fundamental question remains: Why were the overwintering Blackcaps traveling north in winter doing so well? "Now that is just the sort of question we can answer using the Garden BirdWatch data!" exclaims Kate Plummer. The first task was to verify the apparent increase in the numbers of overwintering Blackcaps. Kate extracted the Blackcap data from twelve consecutive winters (1998/99 to 2010/11) from almost 4000 separate locations, amounting to a total of over 800,000 weekly records. While bird-watchers have been talking about the increasing numbers for some time, Kate's careful analyses show that the increase has been particularly pronounced since the early 1980s, with numbers peaking each year in late January and February.[21] Currently, about a third of all gardens in the program have Blackcaps. But why?

Well, because of food availability obviously, but as there does not seem to have been a major change in natural food supply, feeders clearly come into the picture. But what do Blackcaps like in the way of feeder fare? As it seemed likely that the answer to this question was fueling—literally—the Blackcap boom, Kate Plummer decided to dig deeper. Garden BirdWatch participants who observed winter Blackcaps regularly were invited to report on which types of feeder foods the birds were using. With over 500 people sending in data from all over the country, this was a decent sample of the diversity of feeder supplies. The results showed clearly that Blackcaps preferred fat and sunflower hearts above all else, and that the presence of the birds in gardens was strongly associated with the availability of these foods.[22] "This move into gardens seems to be becoming stronger over time," Kate explains. "It seems to suggest that Blackcaps are progressively adapting to obtaining their food from feeders." This idea appears to be supported by recent studies indicating that the overwintering Blackcaps have developed a narrower and stronger bill than their traditional cousins, possibly indicative of a more generalist diet.[23] Interestingly, many people also noted that this otherwise modest little bird is highly aggressive at feeders, vigorously chasing away other species attempting to use "their" feeder.

Kate Plummer's well-reasoned conclusion is that the general availability of these foods at feeders over broad areas has provided the foundation for the growth in the overwintering Blackcap population in Britain. The period during which numbers of the species has steadily increased—starting around the 1960s—coincides closely with the growth in popularity of bird feeding and the proliferation of commercial bird foods, including all manner of fatty products. Although it will be difficult to prove conclusively, the advent of the Blackcap as a common winter visitor to British gardens may be one of the clearest examples of the direct influence of feeders on the abundance and distribution of a species.

But wait, surely this period also coincides with the general warming of the climate, a well-documented global phenomenon with a wide range of ecological influences. This would have taken some of the edge off the late winter bite for the birds arriving from Europe and marginally improved their chances of surviving.[24] The critical element to this situation is, of course, the availability of suitable foods. Blackcaps are known to depend largely on the fruit and berries found in woods in winter, moving into gardens when these supplies are exhausted. The somewhat milder conditions associated with climate change probably enhance survival rates, but only if appropriate foods are available. Warmer but still hungry just won't work.

The "Blackcaps in winter" story is compelling but it is also exasperating. It is probably one of the clearest examples of the effect that feeders may be having on a species. But these insights are only possible because we just happened to know a lot about this particular species, and that is because scientists were already focused on other aspects of Blackcap biology—their migration patterns, reproduction, and evolution—and because bird-watchers were gathering data at their feeders in sufficient detail that critical questions could be asked. It is exasperating because despite the staggering amounts of excellent data being garnered by programs such as Garden BirdWatch and Project FeederWatch, there are so few clear examples of the feeder effect.

My time spent discussing the magnitude and limitations of the grand feeder-focused programs with people from the Cornell Lab of Ornithology and the British Trust for Ornithology was also characterized by the possibility of extraordinary opportunities. Both organizations emphasized that while they themselves were frustrated in having too little time to engage in interrogating the information they were so successful in attracting, this data was available to anyone with legitimate interests in collaboration. "It's a gold mine for the right kind of curiosity," reiterated David Bonter, as I was leaving. "There are answers here for questions we haven't even thought of yet!"

This particular excursion has been an attempt to see if it was possible to discern specific ways in which providing food for wild birds in our gardens was actually influencing them in some way. We tend to assume that our private feeding activities are either neutral or positive, if we think about the consequences at all. By now, however, we should be aware that the scale of our collective efforts in supplying vast amounts of bird food over large areas is potentially of enormous impact. Certainly, there are some very strong suggestions—hummingbirds in wintry Ontario and Seattle, the recovery of House Finches, Blackcaps wintering in Britain then breeding early in Germany—but establishing a direct and definite link between bird feeding and such changes is always going to be difficult.

Seeking to understand the possible relationship between the practice of feeding and the outcomes for the birds themselves is perhaps the central objective of this entire journey. The partial glimpses of possible explanations—at best—described above have led me to reconsider some of the more fundamental aspects associated with why we feed birds in the first place. It may seem trivial and even redundant, but I think it is worth asking some of the basic questions again and seeing how much we really know. Bear with me.

Why Do We Feed in Winter?

Well, that is about as fundamental as it gets, but let's make sure we are clear about why birds are fed in winter in the first place. Although aspects of this question will be dealt with in some detail in later discussions of the motivations for feeding (Chapter 7), the key points need to be considered here.

First and most obviously, winter feeding seems to be about assisting the vulnerable, a humane act to help creatures in need. Although there is certainly an element of birds representing "life in the depth of the apparently life-less winter" (as one correspondent described seeing hummingbirds at a snow-covered feeder), "helping the helpless" seems hardly contestable as a primary motivation. This assistance can have two main forms: providing much-needed sustenance for apparently desperate birds due to current severe conditions; and second, providing additional food that improves their chances of surviving through to the next season. If you like, there is the immediate relief of hardship—an act of welfare—and the longer-term goal of preservation, by prolonging life. For most people, I suspect that these components are not really considered separate: we are simply providing care, now, and hoping that this makes some sort of difference in the longer term. This is pretty straightforward: the primary goal of feeding wild birds in winter is actually about life and death.

So, what is the evidence to support these straightforward claims? Evaluation of the evidence should be done in the dispassionate manner of the apparently objective scientific method, although that is all well and good in theory. For many people, including myself, being passionate about bird feeding is why we are engaged in the activity, why we are feeders in the first place as well as in the long run. But for me, this also means that we need to ask serious questions and assess the best available evidence. It does not mean searching high and low for data that support our position or refute opposing views. It does mean, however, a willingness to be open-minded. With this in mind, we say with certainty that the main apparent benefits of winter feeding are short-term welfare and longer-term survival. But does winter feeding achieve what is claimed?

As will already be familiar, the amount of reliable information available is quite small. There are some quite obvious reasons for this dearth of relevant information: it is very difficult to obtain. Being able to assess whether having access to feeders assists welfare or enhances survival requires detailed information on a large sample of individual birds, which enables any of the possible changes to be monitored. Two key features that might be examined are changes in body weight and the probability of surviving over a certain period of time. Obtaining these sort of data requires, at least, catching—without harm—lots of birds and taking extremely accurate weight measurements. This is hard enough, but when it involves mist-netting tiny birds such as chickadees and tits that weigh only about 12 grams (half an ounce) in the depths of winter, the challenges are formidable.

Thankfully, some such information is available. Working at the northern limit of the tiny but hardy Black-capped Chickadees in the cold winters of Wisconsin, Margaret Brittingham and Stanley Temple captured and color-banded (ringed) over 500 birds and followed their fortunes over three tough years.[25] These birds were not, however, suburban chickadees visiting who-knows-how-many garden feeders. To be certain of the effects of specific amounts of seed, these researchers moved out of town into woodland study sites where they could control the supply of food—at least the amount provided by humans—available to the birds. They had to assume that the amount of natural food was the same in the various sites they worked in. To really understand the influence of the provisions, they had to be able to compare chickadees with access to feeders with those missing out. In other words, Brittingham and Temple did not just observe what was happening, they manipulated the key feature by setting up an experiment, providing food (in this case, sunflower seeds) in one place and not in another. The importance of such "supplementary feeding experiments" in enabling us to assess the effects of bird feeding cannot be overstated; the entire next chapter is devoted to reviewing this type of research. Here, however, we are focused on specific questions raised about winter feeding.

So what difference did feeders make to these chickadees in winter? The careful weights recorded showed that, on average, the fed birds were significantly heavier than the unfed birds. Although this difference was only 0.13 grams (0.004 ounces), or just 1% of adult body weight, it represented about two hours' worth of foraging time during the coldest periods. Similar results have also been obtained for various species of tits in Europe:[26] the additional food made a major difference when the conditions were really cold. At other times or in places where the winters are relatively mild, the birds were less likely to feed so regularly at feeders, finding most of their dietary needs in the wild as natural food. During extreme conditions, however, the seemingly modest differences in weight translated directly into the ability to survive: birds using feeders were much more likely to stay alive into the following month compared to unfed birds. Most clearly, the Wisconsin chickadees with access to feeders were—on average—twice as likely to make it to the next winter, a major benefit for a bird that lives only for a few years.

Okay, these are important findings and certainly provide good evidence in support of the main claims for winter feeding. Nonetheless, we need to be clear about what they do and do not say. First, while these studies are sound and reliable, they are limited to several closely related species (tits and chickadees belong to the same family), and while these are among the most abundant feeder birds in the Northern Hemisphere, we cannot simply assume that the effects will be similar for other species. There are numerous survival studies on a range of birds but, remarkably, none on the garden birds that use feeders. Even more important, the very conditions of these carefully conducted studies—strict control of the amount, type, and timing of the food, ensuring that the subjects don't have access to other feeders, for example—essential for scientific rigor, mean that these experiments may bear little resemblance to the actual world of a bunch of birds in a suburban backyard. A typical garden-variety chickadee or tit may be able to visit several feeders, as well as nearby parkland, woods, or school yards. They may be at risk from cars and cats rather than hawks or owls, and may have been raised in an environment full of people and their edible refuse. Most wild bird feeding occurs in anthropogenic landscapes—towns and cities—and we need to bear this in mind.

Should We Be Feeding All Year Round?

There is, therefore, at least some support for the idea that feeding birds in winter assists their survival. While the evidence is limited to a few species, this effect does seem fairly logical and sensible (we will deal with these issues in more detail in the next chapter). But what about the equally intuitive reasons given for why feeding should be limited to winter? This issue was covered quite extensively in Chapter 3, but we are revisiting some of the fundamental claims and may need to be reminded that bird feeding was fairly strictly winter-only in most places until quite recently.

Perhaps the best place to start this part of the discussion is back in the countries where the norm for feeding is still unequivocally restricted to winter. Whether through informal cultural lore or advice from reputable organizations, most people feeding birds in these countries seem to be well aware of why feeding birds should be limited to the cold weather. The Estonian Ornithological Society (to pick a less familiar example), for instance, advises that (my translation): "Birds do not need extra food from people" except during the prolonged cold of winter. Then, however "it is a big help for those species that have decided to stay," but "bird feeding must be completed no later than April."[27]

Two of the most commonly promoted concerns are, first, the possibility that the birds may become dependent on human-provided foods; and second, the risk that breeding birds may bring inappropriate feeder foods to their nestlings. The official guidelines from countries the world over are (or were) full of pleas for feeders to ensure that birds be left to "look after themselves" when not threatened by severe weather, as well as disturbing stories of chicks choking on a peanut.[28] In the 1980s, for instance, the Royal Society for the Protection of Birds in the UK stated that birds "should not be allowed to become dependent on the provider" and "that nestlings may be killed when fed with indigestible foodstuffs."[29] There are also emerging concerns about the nutritional influences of typical high-fat winter foods, but these are more recent issues (and are considered in some detail in the next chapter). For the moment, let's briefly look at what is known about dependency and chicks being fed the wrong foods.

Do Birds Rely on Our Food?

The possibility that the birds we feed may become dependent on the food we provide has been shown to be a central concern to feeders everywhere. It is also among the first issues raised by both opponents and proponents of feeding as reported in a wide range of surveys from throughout the world.[30] Obviously, as a passionate bird-watcher and feeder, I need to be comfortable that my pastime is not causing harm, but also that is not changing the natural balance too much. If I was to learn that "my birds" relied entirely on my remembering to fill the feeder, I would be quite concerned. Even more graphically, many of the guidelines on bird feeding strongly intimate that birds could become so used to finding their food at the feeder that they could "forget" how to forage naturally.[31] This would result in (cutting straight to the worst-case scenario) starvation. Hence, the oft-cited "golden rule": once you start, don't stop![32] Indeed, this may be among the most widespread and widely accepted of all tips on feeding. And it is worth stating things as starkly as this because I am very much aware that this is indeed the situation many individuals now find themselves in. There really are large numbers of people who feel they cannot stop feeding—ever—because of the potential catastrophe that would surely result. This is the very reason given for having a strict limit to how long feeding should last. In other words (it is argued), we should feed to assist the birds during the really cold periods but stop as soon as possible in order to limit this possibility of dependency.

This issue was of such importance to my own studies of bird feeding that for many years I have attempted to gather all the relevant studies and to speak to as many people as possible from around the world who have thought about this seriously. Surely something so salient to the practice of bird feeding must have been subjected to critical scientific attention. The reality is, again, that there has been extremely limited research on this issue. But, to be honest, this is difficult research to do, especially the kind of experiments that would be necessary to actually test whether birds really were dependent on feeder food. Theoretically, that would require making the tough call of providing food to birds, then stopping the feeding entirely and observing carefully what happened next. Could anyone actually do such a thing?

Well, yes. In a simple but game-changing study, Margaret Brittingham and Stanley Temple again show why they have been among the most important pioneers in the scientific study of bird feeding.[33] Focusing directly on the core issue of dependency, these researchers decided to see what might happen if birds—chickadees—were deprived of their feeder food (sunflowers) over a whole winter. What makes this work so compelling is that the site where the feeding was stopped had been providing the local birds with supplementary food for 25 consecutive winters! If any birds were going to be relying heavily on feeders, this was them. To be certain of the outcome, Brittingham and Temple set up a separate though identical site where the chickadees had never been fed. Both sites were surveyed weekly over three winters to check on the presence and behavior of individuals. This point—that the researchers were able to track the fortunes of specific birds—is the main difference from the earlier study by these researchers mentioned previously (which assessed general survival risk at the population level). This new study was focused on individual chickadees, including observing their foraging activities, especially during the third winter, during which no feeding occurred.

The result was spectacularly dull: there was no significant difference in monthly survival rate of individual chickadees at the two sites. Most significantly, the lack of supplementary food made no apparent difference to birds who had previously had access to unlimited seed. How on earth could they have survived? Although they do not say as much, it is clear from the way their paper is written that Brittingham and Temple were not at all surprised by the results. Their intimate knowledge of these otherwise delicate and vulnerable birds indicated that, even in the depth of a snowy winter, the birds were able to find most of their food from natural sources. The researchers observed their birds feeding on a wide range of items including mites, insect eggs, and larvae, as well as grain gleaned from horse dung. The sunflower seeds available at the experimental feeders were just one source of food among many. Indeed, even during the first two years of the experiment, when the feeders were still being stocked, almost 80% of the birds' diet remained natural. Addressing directly one of the major concerns associated with the issue of dependency, the researchers state: "It is not surprising [after 25 years of feeding] that these chickadees had not lost their ability to utilize natural food efficiently."[34] Even these hard-pressed Wisconsin birds treated the food supplied by people as, well, supplementary.

This is crucial work, but even so, the researchers also caution against applying their findings too broadly. Theirs was, after all, a carefully controlled experiment in a natural woodland; the conditions in a typical suburban garden may be vastly different, especially in relation to the availability of natural foods. The caveat is especially critical in the case of places were the foods provided at feeders may actually represent most or all of the foraging resource present, as is likely the situation where once migratory species now overwinter apparently because of reliable feeders or when climate conditions are pronounced or prolonged. These are obviously situations where continuing to feed in winter is vital. But these are also perhaps best appreciated as appropriate responses to local conditions. The middle of winter is one thing; at other times, provisioning may not be so significant in the lives of the birds. (Note that I did not suggest it was not significant in the lives of the people doing the feeding.) What the Brittingham and Temple study more than hints at is that for most birds, most of the time, our feeders probably provide only a portion of their overall diet.

A couple of more recent experiments are also worth mentioning here: the first looked at what happens in relation to a whole community of birds denied access to their feeders, while the second "withdrawal of feeders" study was actually conducted in the suburbs. Travis Wilcoxen and a large group of colleagues undertook a particularly important feeding experiment in woodland sites in Illinois where they compared a lot of health features of many species with and without feeders.[35] Some of these results will be described later, but here we can mention what happened when the feeders were removed after two years. Although some of the positive features associated with having a regular feeding source—higher fat levels, increased antioxidants (see below)—disappeared when the feeders did, for other characteristics there were no differences with feeder-less birds. Significantly, these researchers concluded that the feeders were indeed "supplementary" to the birds' diet.

As part of her pioneering work undertaken in Auckland, New Zealand, Josie Galbraith asked lots of people to let her set up her purpose-built feeders in their gardens so that she could record the activities of the birds.[36] Then she had to convince about half of the willing participants to stop feeding, 18 months after they had started! Yet apparently they did so, because the huge increase in the abundance of the common species that occurred while the feeders were available simply dissipated when the food stopped coming. The concentration of birds at the feeders went back to typical prefeeding numbers, with the birds apparently adjusting to the change in resources. Again, this seems to suggest that birds are more than capable of altering their foraging behavior to differing conditions, even in the apparently "artificial" suburban environment.

This was an issue that I was forced to take very seriously during my own research into the foraging of Australian Magpies living in the suburbs of Brisbane, Australia.[37] Aspects of my work are described elsewhere in this book, but the key point was that although we made detailed observations of the natural foods eaten by the birds while they were in view, we really don't know what they were getting up to when they were out of sight. The same is true for anyone wanting to understand the actual diet of birds in human-dominated landscapes: we have little idea of what they eat and where they go when they leave the place where we were watching them. In the early days of our magpie studies we first became aware that feeding stations for these birds were everywhere! Indeed, they were so common, offering a smorgasbord of meats and cheese, we wondered whether they presented an opportunity for magpies to abandon natural foraging altogether. With so much high-quality protein, so easy to find, why should the birds bother digging up worms at all? Furthermore, we reasoned, this would be even more likely when the adults had a nest full of large, hungry nestlings. If any birds were set up to become reliant on a source of supplementary foods, it was surely breeding magpies.

With this logic in mind, Rebecca O'Leary and I set out to compare the source of the foods parent magpies were feeding to their chicks in territories containing feeding stations and those without.[38] Actually finding places where people did not feed magpies proved remarkably difficult. Because suburban magpies have quite compact territories, however, we eventually found similar numbers of magpie pairs that were fed and not fed for comparison. We (well, mainly Rebecca) then got out the telescope and began to carefully document what the parents were bringing to the nestlings.

To our great surprise, almost all the baby food was entirely natural—worms and beetle larvae—for both the fed and unfed pairs. The really notable thing was that the parent magpies were mostly ignoring the readily available meat-laden feeding stations located all around as they searched for good places to probe for grubs in the ground. We concluded that these parent magpies seemed to have a "no junk food" policy when it came to providing for their nestlings.

This was also a conclusion reached (but not known to us at the time) by Richard Cowie and Shelley Hinsley in the early 1980s.[39] These pioneering bird-feeding researchers compared the diets of tits in rural and suburban sites around Cardiff, Wales, and learned that despite the abundance of feeders in the suburbs, both groups of chicks were fed high proportions—over 85%—of insects, although some seed was also used regularly. I have discussed this finding with colleagues from around the world, and it turns out to be a familiar story: parent birds of many well-studied species typically bypass all those feeders as they seek out the insect food they seem to "know" their chicks need. This was another reassuring finding and unexpectedly addresses the two concerns we have been discussing: Does feeding lead to either dependency or inappropriate provisioning of nestlings? In the few cases available, the answer appears to be "no" in both cases. Provisionally, then, we feeders can relax a little; the birds seem to know what they are doing. Thankfully.

Provisionally? Yes, of course. As we would all be aware, there are some situations when withdrawing food supplies would certainly be hazardous and, indeed, unethical. Prolonged periods of extreme weather such as blizzards, storms, or severe drought all present serious challenges for birds seeking sustenance. Similarly, the increasingly frequent instances of certain species apparently overwintering instead of migrating—perhaps because of the reliable availability of feeders—may represent genuine reliance. In these cases, it is quite likely that the birds would simply die without our providing food.

Some have argued that providing food in these situations is perpetuating an artificial regime, that we are interfering with the processes of natural selection. Others, while acknowledging that to be possible, nonetheless counter with the welfare argument: that we bear clear responsibility for the well-being of the birds that have chosen to utilize the resources we have provided.[40] And yet others suggest . . . a range of philosophical positions, which we will return to later in this journey. For now, we can accept that for most of the time, our supplies are just a part of the birds' overall diet, although there are also certain times when external conditions are such that we really do need to keep feeding.

Are Feeder Items Used as Food for Nestlings?

This discussion returns us to the important issue of parent birds utilizing inappropriate feeder items as nestling food. As we have seen, most observations seem to indicate that breeding birds typically do not use feeders as a significant source of food for their young. Although some studies of tits did find that some seed was used in this way, invertebrate foods—such as caterpillars, spiders, aphids—were always the most common items.[41] A more specific concern, however, is the risk of choking on a large item such as a peanut, an entirely valid worry. Thankfully, this has been rarely observed; nonetheless, it does happen. Richard Cowie and Shelley Hinsley seem to have been among the first to report dead Great Tit nestlings (only three) with large pieces of peanut stuck in their tiny throats,[42] a story repeated endlessly ever since. Other researchers have related similar stories in other species—including Florida Scrub Jays—but always as rare and atypical events.[43] These incidents really do appear to be the exceptions that prove the "seed is not used very often as baby food" rule. Where it did occur, there were strong suspicions that the particular birds involved were inexperienced or, more commonly, simply desperate. After all, we now know that large numbers of all tit nestlings in suburban environments starve before they leave the nest, almost certainly because of a lack of suitable insect food.[44] Feeding seeds to chicks may actually be evidence of the general lack of suitable insect food in the local area and of the grim reality for the parents valiantly attempting to keep their brood going. Feeding seeds to chicks may be a symptom of poor conditions but is unlikely to be the main cause of the breeding failures. Significantly, Cowie and Hinsley and others have found no relationship between the amount of seed in the diet of nestlings and their likelihood of survival, another reassuring result.

Does Feeding Change Bird Populations and Communities?

Finally, we need to consider perhaps the grandest of all the claims associated with feeding: that the provision of feeders actually changes the numbers of birds being fed. This is crucial because such an effect underlines some very important propositions: that feeding aids the conservation of declining species, and that feeding may increase the abundance of pest, invasive, or just unwelcome species. How much do we know about these important issues?

The global popularity of wild bird feeding means, among other things, that a huge amount of food is offered—and accepted—by birds throughout the world every day. Various figures have already been cited to emphasize the volume of this provisioning—for instance, 60,000 tons of bird seed is distributed annually in the UK, or enough to support 71 million birds (twice the estimated population size of the top ten feeder species!).[45] Another metric sometimes mentioned is the number of feeding devices being presented. One study from Michigan found a density of 8.5 feeders per hectare in urban areas, 2.6 for suburban, and 2.8 feeders for rural areas. This averages out at about 600 bird feeders per square kilometer (or twenty-three per square mile) for this state.[46] Similarly, in the UK one national survey found the density of feeders to vary from about fifty to over 1000 per square kilometer (twenty to 390 per square mile) with an average of 925.[47] It probably doesn't mean much—given that we don't have similar data for almost anywhere else—but one location within the city of Sheffield in England was found to have the highest density of bird feeders in world: 1208 per square kilometer (460 per square mile).[48]

Beyond the simple fact that these are apparently large numbers, and that they indicate something quantitative about the popularity of bird feeding in these different places, we also need to acknowledge that these figures represent a truly massive intervention by humans in the lives of birds. If food is a fundamental influence on the lives of the birds around us, we are almost certainly changing things in some way. How much do we really know about these possible changes?

The answer to that question is "frustratingly little!" It is surprising, but despite the enormous numbers of people actively engaged in bird feeding at their homes, the growing scientific interest in urban environments (as indicated by the emergence of the field of "urban ecology"), and the scale of the bird food industry itself, our knowledge of the effects of wild bird feeding is quite meager. Let me clarify this claim. We do know a lot about the preferences many species have for the different types of food, both commercial and otherwise, and much research has been conducted into the nutritional and dietary balances and suitability of many seed mixes and bird food products. The results of these detailed research programs mean that—with a moderate level of care—we can be reassured that the items we buy to be offered as food are likely to attract the species we wish to attract and that the food value of these products is suitable.

These are research topics of primary interest to the massive bird-food industry: the suppliers, the seed companies and the increasingly discerning consumer/feeders. If the birds don't eat some fancy new mix, there are plenty of other products to choose from. This is the normal dynamic of the commercial marketplace and has led to the proliferation of products and gadgets we see today. What is still missing, however, is a clear understanding of just what happens when birds are attracted to our feeders. While we may enjoy seeing lots of chickadees or tits, what does it mean for the birds themselves?

There has been, thankfully, one very recent and important study that does provide at least some light here. Led by Travis Wilcoxen (mentioned above) the study assessed a wide range of physical features of the community of birds using their experimental feeders in woodland sites in Illinois and compared them to the features of birds without access to feeders.[49] The feeders supplied exactly the sort of general seed selection typical of the area and were replenished three times per week for two years. Over 1600 birds of 11 species were carefully captured and examined in one of the most comprehensive and relevant studies so far conducted. What they found was genuinely reassuring: in general, the fed birds scored significantly better on numerous parameters than the unfed birds: increased antioxidants (an indicator of health), reduced physiological stress levels, more rapid feather growth, and generally better body condition. Overall, this was the first decent evidence suggesting that feeding might actually be good for the bird's health.

But of course, there was a down side too. The feeder birds were also much more likely to show signs of several diseases (especially avian pox)—exactly as expected from a process that brings lots of birds of different species together—although there was also some evidence that sick birds with access to feeders were able to recover more rapidly. So, some good news mixed with some (not unexpected) bad. More studies like this are desperately needed.

Is More Less?

The density of feeders, a measure of the anthropogenic feeding resource available, provides one measure of how the distribution of bird food may be influencing the numbers and diversity of species of birds across the landscape. Because the number of feeders per area varies from none to lots, it is possible to assess whether spatial changes in food availability relates to differences in bird numbers. Two British studies lead by Rich Fuller, conducted at both national and local levels, provide clear evidence that feeder density strongly predicts bird abundance: the higher the density of feeders, the more birds.[50] But lots of food did not mean more species, just more individuals. The second study found much the same pattern, though with additional details of certain species. In Sheffield, at least, the density of House Sparrows, Blackbirds, and Starlings related directly to food supply. Somewhat surprisingly, this relationship did not hold for either Great or Blue Tits, despite them being feeder stalwarts. The same pattern was found in Michigan, with more feeders leading to more birds in general but not more species.[51]

This general pattern of an increase in food supply leading to a larger abundance of birds—more individuals—but without attracting additional species is a familiar picture to ecologists studying birds in cities. Compared to forest or woodland, urban environments across the globe exhibit the same pattern of having far greater numbers of birds, particularly pigeons and corvids, but distinctly lower diversity of species.[52] Although there are many possible reasons for this pattern, including fewer predators, warmer temperatures, and the tendency for some species to cope better with the presence of humans, by far the greatest influence seems to be the foraging opportunities provided by cities. In their review of how birds living in cities have been affected, Dan Chamberlain and his colleagues concluded that "food availability was paramount," with human-provided foods being the primary source.[53] Rich Fuller and his colleagues went so far as titling one paper: "Garden bird feeding predicts the structure of urban avian assemblages."[54] In other words, all those feeders are definitely having a fundamental influence on the composition of bird life in our cities.

Of course, simply having more birds may be a worthy goal in itself. At the level of one's own backyard, attracting a colorful congregation of favored species by providing feeders may be exactly the intention. It is important to appreciate, however, that a reliable and plentiful source of food represents an artificial concentration of a high-quality resource that will certainly attract many birds. Ignoring for the moment the other possible implications of bringing lots of birds, and species, into close contact at a feeder (increased aggression, attraction of predators, exchange of disease, for example), simply seeing lots of birds does not necessarily mean a larger local population of that species.

This was certainly the case in Auckland. Josie Galbraith's innovative work on bird feeding in the suburbs of New Zealand's largest city has already been mentioned above in the context of understanding what happens to the birds when feeding stops. While this earlier study found that the birds simply dissipated to preseason levels, her more recent research investigated how the initiation of feeding influences the overall community of birds.[55] This work is especially important because it appears to be the first to investigate the feeder effect in places where feeding had not occurred previously. By comparing the diversity and abundance of birds before feeding started to what occurred 18 months later, Josie was able to demonstrate significant increases in the numbers of most of the already common species but virtually no change in the number of species. What was of most interest, however, was which species prospered or and which did not. In Auckland, feeding greatly increased the numbers of House Sparrows, Spotted Doves, and Starlings, all species introduced to New Zealand. House Sparrows, already the most abundant bird prior to feeding, more than doubled its numbers with feeding. Although numerous native species were present, feeding affected only one, the Grey Warbler, and this was to decrease their numbers.

Josie interpreted the changes to mean that the type of foods offered has fundamental importance. Her experimental feeding regime replicated what most New Zealand feeders were offering—bread and small seeds—and this almost certainly favored the foreign species, famously granivores or eat-anything generalists. The local natives, such as White-eyes (Silvereyes), Tuis, and Grey Warblers, on the other hand, were nectarivores or insectivores; those bird tables would have been of little interest. For a place like New Zealand, desperately trying to maintain its unique wildlife in the face of an ongoing insurgence of foreign invaders, these are sobering though illuminating findings. Time for some feeding advice?

Special Seeds

Sometimes, however, there is a very clear influence of a particular food type on the abundance of certain birds. Of course, this was what happened when black sunflower seeds first began to be provided widely: the seed was spectacularly popular with a wide variety of species, with birds flocking to feeders everywhere. This was probably the single biggest influence associated with bird feeding, but was so generalized and widespread it was impossible to document. By the time researchers began to take an interest, the phenomenon had stabilized. We simply cannot know just what the bird community would look like without black sunflowers because they are effectively already everywhere.

It is, however, possible to know more about the likely impact of more recent seed introductions. The clearest example would probably be that of the small black high-oil seed known as nyger (or nyjer or niger, and sometimes "thistle"). Although this seed was known to be highly attractive to American Goldfinches and Pine Siskins in North America as far back as the 1960s, it was the remarkable reaction of Goldfinches to what was a new offering in Britain that was quickly noticed.[56] Goldfinches are one of the European species that traditionally were seen only in gardens when supplies of their natural diet—the seeds of dandelions, ragworts and groundsels, for example—were at a low ebb. The arrival of nyger changed all that virtually overnight, according to Chris Whittles, founder of CJ Wildbird in the UK. "Nyger was a niche product mainly used by cage bird enthusiasts as a 'conditioner' prior to breeding. The wild goldies, but also Lesser Redpolls, Siskins, and Dunnocks to some extent, also seem to really go for it just before the breeding season. But the way that the Goldfinches took to this one seed type as soon [from the early 1990s in the UK] as it was available was phenomenal." The BTO's Garden BirdWatch data for Goldfinches matches this observation: over a 12-year period, Goldfinch sightings went from 23% to 60% of all gardens.[57] That's a lot of birds. Of course, these reports were from gardens and not the countryside, but it does seem that this is one species that is actually increasing in abundance nationally in Britain. "It's hard to prove, of course," cautions Kate Risely from BTO. "But the growth in Goldfinch numbers does correlate fairly well with the advent of nyger."

So it is possible that this may be a case where effectively the entire national population of a particular species really has benefited from a particular feeder food. Another example is almost certainly going to be a surprise, unless, of course, you live in the city of Reading, a little west of London. Here a most unexpected bird is being actively encouraged to visit local gardens by the provision of an unusual type of bird food: meat. Large amounts of pork, beef, and lamb, but primarily chicken, are being placed in gardens with the explicit aim of attracting a large bird of prey, the Red Kite. Now many feeders have an uneasy relationship with their local birds of prey, particularly species such as Sharp-shinned Hawks or Sparrowhawks that are often found in urban environments. Indeed, considerable efforts are made to minimize the risk to smaller birds visiting garden feeders from avian predators. Yet according to recent research by Melanie Orros and Mark Fellowes, almost all of the people feeding Red Kites in Reading also fed smaller species.[58] "They did worry about their smaller garden birds," Mel Orros told me when I visited her, Red Kites wheeling over the nearby M4 motorway to London. "But the wonder of seeing such a majestic and wild creature swooping down into your own back yard is really a special experience, especially in a typical English town." "Of course," she adds, "they also believe that they are assisting in the recovery of a previously extinct species!"

Reading happens to be relatively close to the Chiltern Hills, where Red Kites were originally reintroduced in 1989, having been absent from Britain for centuries. The remarkable story of the return of a large raptor to the countryside—largely through the use of supplementary provisioning of vast amounts of meat—is described in detail in Chapter 8 as a particularly successful example of feeding for conservation. For our present purpose, however, this background provides a valuable context to what may seem an incongruous activity. Many local people were well aware of the reintroduction program and were pleased to know that a great conservation story was unfolding nearby. But when Red Kites started appearing within the city itself, the possibility of being able to see Red Kites close by in one's own garden was certainly worth a try. Mel showed me astonishing photos of these fearsome-looking birds of prey swooping down to pluck bits of meat carefully arranged on the back lawn. I couldn't help wondering about whether the Blue Tits would still visit the feeder in the background.

Benefiting the Wrong Species

While we all have favorite species, including those that turn up only occasionally, there are also birds we would much rather not attract. This may be because these species are unwilling to share the feeder, are particularly messy or noisy, or simply because we don't like them for no particularly obvious reason. Sometimes these species may be overtly aggressive and their presence may prevent any other birds from visiting; there are plenty of stories of people who have given up feeding altogether because they attracted only species generally disapproved of: various corvids, grackles, Starlings, and feral pigeons all come to mind. More common are situations where feeders attract large numbers of certain species whose presence prevents other species from visiting, or invasive species such as Common Mynas, exotic pigeons, and (in places where they are not native) sparrows. In the main cities in New Zealand, for example, the most common feeder birds by far are House Sparrows and Blackbirds, both introduced to that country and doing very well indeed.[59] As has been known for some time, some of the "cheap" seed mixes are notorious for attracting a range of unpopular species, possibly because part of the success of these birds in urban environments has been their willingness to try a wide range of foods. Again, there can be little doubt that the abundance of these familiar "street" birds has been at least partially due the presence of so many feeders.

A particularly spectacular example of this relationship between feeders and undesirable urban birds is that of two most unusual invaders: tropical parrots now living in the Northern Hemisphere. Feral populations of both Monk Parakeets and Rose-ringed Parakeets (often called "Ring-necks") originated from aviary escapees and both are thriving in their extremely different new habitat. The Monk Parakeet, a native of Argentina, now lives in highly social (and noisy!) aggregations in numerous locations in the eastern United States but most famously in the wealthier suburbs of Chicago. A study of their diet in this area found the expectedly broad and opportunistic diet for most the year, but with the coming of winter, these birds relied entirely on local bird feeders for all their food.[60] In this location, Monk Parakeets persist only because of feeders, which are definitely not being provisioned with them in mind.

It's a similar story for the Rose-ringed Parakeets now spreading across several European countries including Spain, Germany, and Belgium.[61] Their numbers in southern England are also rising steadily, and large, conspicuous flocks of these exotic-looking parrots are now regularly spotted around the picturesque parks of inner London. While subsisting on a wide range of foods, like the Monk Parakeet, Ring-necks are frequent users of feeders, even managing to use the hanging tube feeders provided for much smaller species. This raises some obvious concerns about the possible influence that these large and assertive parrots could have on the usual birds coming to feeders. After all, being on average about 400 millimeters (16 inches) in length and 120 grams (4.2 ounces) in weight, Rose-ringed Parakeets are about ten times the size of your typical Blue Tit (12 grams, or 4.2 ounces); there would not be much of a contest if these two species were trying to visit a feeder at the same time. Indeed, in some places, the presence of these big parrots appears to have effectively stopped most typical species from visiting feeders. In the center of Antwerp in Belgium, for example, friends who have been feeding birds for years report that since large numbers of parakeets arrived a few years ago, they simply don't see tits in their gardens.

I still vividly remember my first encounter with wild Ring-necked Parakeets. This was not in the dry, hot woodlands of India or Africa where they occur naturally, but one damp, cold May in London's Regent Park. My bird watcher's ears detected an unexpectedly loud, harsh screech, and large shapes swooped incongruously among the foggy elms. My companion Jim Reynolds identified them immediately: "They have even been seen in Scotland but are especially abundant around London and the surrounding counties where their numbers are definitely expanding." We pause to watch a group of about eight scream past, their size impressive even from this distance. "Imagine that lot arriving at your feeder!"

Such an unlikely scenario is now all too familiar to many people in London. There are plenty of stories of these parakeets taking over the nests of other hole-nesters such as Little Owls and Jackdaws. Some have even used that massive beak to destroy wooden feeders and nest boxes. But do they actually affect feeder visits by the smaller species? While Blue Tits and Great Tits, the typical English feeder visitors, obviously would not attempt to compete, they may simply stay clear of the feeder until the larger bird has left. Plenty of species share common areas, more or less amicably, in the natural world, simply by avoiding one another. On the other hand, the presence of these big, dominating birds may simply be overwhelming for the smaller species.

These were issues investigated by Hannah Peck and her colleagues in an ingenious series of experiments undertaken in over forty London gardens.[62] To simulate the presence of parakeets, a tame (but typically noisy) caged parakeet was placed near existing feeders, and the number and behavior of wild birds visiting was recorded by video camera. The information collected was then compared to feeder visits in gardens without parakeets. The really clever part of this research was determining whether the small birds were reacting to the bird itself or whether being aware that parakeets were apparently present nearby was sufficient to worry them. This was simulated by using recordings of their screeches broadcast from speakers hidden in the gardens. Finally, to ensure than any effects were not simply a general reaction of small birds to big ones, a large species occurring naturally in the area, the Great Spotted Woodpecker, was also presented, again in a cage and as a broadcast call.

Despite the apparent complexity of all the components of this experiment, the results were remarkably clear: tits really didn't like coming to feeders if they perceived parakeets to be around. The presence of the parakeet—both the bird itself and also the call alone—led to a consistent and significant reduction in the number of visits to the feeder, and those birds that did visit spent less time actually feeding and more time being vigilant (that is, looking about). The tits did react to the presence of the woodpecker, but far less dramatically. These results are important in being the first to demonstrate clearly that the simple presence of this otherwise nonthreatening invasive species can have a marked influence on the birds coming to feeders.

But, as usual, that is not the whole story. The experiment may have demonstrated that parakeets disrupt feeder visits by tits, but what about the people involved? What do they think? When I met Hannah Peck to discuss her innovative research, she was quick to point out that, contrary to her own expectations, plenty of people were pleased to see these big birds. "I guess I had assumed that all feeders would be concerned about this aggressive foreign species driving out their nice English ones," Hannah explained. "But lots of people were really excited to have these remarkable—and very attractive—parrots coming to their gardens. When you are used to a couple of tits and a robin, the arrival of these vibrant, exotic creatures out of the blue can be a great surprise and treat. It did make me realize that we tend to forget about the people in this story."

And So?

This seems a good point to pause, gather our thoughts, and consider whether we have reached any conclusions about the potential way that millions of feeders may be influencing the birds around us. We have covered a lot of territory—from upstate New York to the suburbs of London—and considered a lot of claims and evidence along the way. This has been intriguing and illuminating, but it has also been sobering to admit that there is still so much we can't be definite about. Certainly, all that birdseed seems to have influenced the behavior of some species, causing them to move to and sometimes stay in places they didn't use to. But it is hard to disentangle this from other factors, such as a warmer climate and changes in vegetation. Feeders seem to have played a role in allowing some diseases to spread, but also in enabling affected birds to survive. And in certain extreme situations, the day-to-day survival of some birds is obviously entirely tied to access to reliable feeders. Yet probably for most birds on most days, feeders are top-ups rather than essentials. If the natural food supplies are available, our feeders are often forgotten.

We don't know what happens when you provide several million tons of sunflower seeds and fat balls every day across entire continents. We can't be sure that seeing lots of Blue Tits or chickadees—or Silvereyes or magpies—is actually a sign of a healthy and resilient ecosystem, or one propped up by artificial foods. We do need to know these things if we are to be entirely satisfied that our pastime is good for the planet and not simply good for us. To tease these issues apart requires some serious science, well-designed experiments that should provide reliable findings. It's time to look at studies of supplementary feeding.

5

What Happens When We Feed?

Insights from Supplementary Feeding Studies

"Maggie" is the inevitable nickname given to innumerable Australian Magpies, even though many are likely to be males. They are big, bold birds, abundant throughout the country, especially within the suburbs where the landscape of scattered, tall trees and endless well-watered lawns provides ideal habitat. These birds are very well known to Australians for two main reasons: their complex and evocative territorial song ("caroling"), which seems to capture the essence of an antipodean spring morning; and, in violent contrast, their aggressive aerial assaults on humans trespassing too close to the nests. It may sound far-fetched, but these attacks are extremely common during the breeding season and lead to lots of injuries and anguish each year.[1] Yet somehow this vigorous combination of art and aggression seems to appeal to Australian sensibilities as the species is often declared our favorite bird. Half the sporting teams in the country seem to be called the Magpies.

Australian Magpies are also well known because they are the most frequent—and favored—visitors to the feeding tables of Australia. This may come as a surprise because while magpies are certainly broad in their dietary tastes (they will eat almost anything), they are primarily insectivores, specializing in worms and grubs gleaned from just below the surface of lawns or grassy fields.[2] They have a similar foraging style to that of American Robins and the Common Blackbirds of Europe, although they are about twice the size. Although Australian Magpies will eat seed and bread and many other things, to attract magpies most people place out a much more carnivorous selection: chopped sausage, ham, pieces of bacon, diced heart, and especially beef mince (ground beef) are all commonly used, as well as cheese and pet foods. The contrast with a typical Northern Hemisphere feeding station could hardly be greater: seed and tiny tits or chickadees versus meat and massive magpies.

The discovery of these veritable butcher's shop smorgasbords throughout the suburbs was an unexpected and significant event in the development of my interest in bird feeding. For some time I have been interested in the wildlife of urban environments and particularly the features that enabled some species to prosper in this strange contrived landscape while many others could not. I was also fascinated by the interactions between people and the otherwise wild creatures they shared the suburbs with. Magpies were an obvious choice for an urban study species: they were abundant, relatively approachable, and most people liked them. And although they had been well studied, to my great surprise this common and familiar bird had not been investigated in its favored habitat, the suburbs. Despite being filled with people (and their cars, cats, and kids), the suburban environment offers plentiful opportunities to an intelligent and resourceful bird. Vast areas of lawns and garden beds for foraging, tall, well-spaced trees for nests and surveillance, and generally fewer aerial predators; no wonder the density of magpies in the suburbs was far higher than in rural landscapes close by.[3] But there was more to the story than simply lawn and trees and fewer hawks.

Food and space are two essentials for survival, and they are closely linked in the lives of many animals. For species capable of defending an area—a territory—against others, a home patch needs to contain all the resources required for both day-to-day survival and reproduction. For the majority of birds, the effort required in defending a patch against determined trespassers and intruders potentially seeking to pilfer your resources or make off with your partner is so demanding that territoriality is typically limited to a few weeks of the year. The process of finding a suitable area, claiming it as your own (usually through vigorous singing and displays), repelling competitors, attracting a mate, building a nest, producing and incubating eggs, and protecting nestlings from weather and predators, among other duties, is exhausting. Most birds, sensibly, are actively territorial only briefly, ceasing their boundary patrols and moving away from the breeding site once the young are mobile.

Not so the Australian Magpie. Unlike virtually all other songbirds, these birds are permanently territorial. This means remaining continuously vigilant for potential intruders and patrolling the boundaries of their territories every day of the year. This highly unusual feature of magpie life has numerous implications for the birds, including the reality that finding a life partner means *really* settling down: remaining in the same patch for the rest of their lives. For such a long-lived species (Australian Magpies frequently live for more than 20 years, much longer than a robin, for example), this makes the resources provided within the territory of critical importance: everything needed for survival and reproduction—and shelter and everything else—must be suitable and sufficient more or less permanently. Although we now know that both males and females do make the occasional furtive and apparently secret sojourns out of their territories,[4] these escapades are almost always brief; in general, most mated magpies spend most of their lives in the same patch. Given the obvious necessities required, it is not surprising that magpie territories in the suburbs are relatively small compared to that of an out-of-town magpie.[5]

This was the background to one of our first studies of the ecology of suburban magpies. There were plenty of worthy questions to pursue, but we chose to investigate what was fundamental to them all: What was the food supply of an average territory in the suburban landscape? With my colleagues Tom Nealson and Dan Rollinson, we began the usual process of modern wildlife research: the careful capture of birds so as to measure, weigh, and attach colored leg bands to allow us recognize individuals, and the methodical observations of the birds as they went about their normal lives. These are activities familiar to field biologists studying birds anywhere in the world, except for the fact that we were attempting to conduct our research in the urban environment. And "urban" means people. Plenty of folks came over to see what we were up to. From simple curiosity and genuine interest to skepticism and outright alarm at our actual or imagined methods,—we spent a lot of time patiently explaining and listening. But we also managed to obtain some important observations of our magpies.

As has been understood for some time, Australian Magpies (like Common Blackbirds and American Robins) use their ears as much as their eyes to hunt for the grubs and worms that make up the bulk of their diet. Through some remarkable experiments using recordings of these invertebrates moving through soil,[6] we now know that those characteristic side-to-side movements of the bird's head as they concentrate on a spot on the ground immediately in front allow them to accurately locate their prey beneath the surface with pinpoint accuracy. A dramatic pause, a sudden thrust deep into the sward, and the next moment a large worm is thrashing at the end of the bill. It's an approach to foraging that predisposes birds such as magpies to success in a landscape dominated by lawns and grassy fields. Magpies do very well because these well-watered, fertilized, mowed, bright-green landscapes, so characteristic of suburbs throughout the world, are also ideal for the invertebrates they thrive on.

Our careful observations of the foraging behavior quickly allowed us to form a clear impression of the richness of lawns and sports fields as a food source for magpies. We were able to compare their feeding activities on lawns versus fields, in suburbs versus on farms, at dry times versus wet, and their foraging behavior for their own consumption compared to when they were feeding hungry young. We felt we were beginning to understand the relationship between the availability of natural foods and the success of the species in this human-dominated environment. It was when we started to focus on the chicks in the nest, however, that we began to realize that we were dealing with only part of the story.

Baby birds still in the nest ("nestlings") are, obviously, entirely reliant on their parents for all their food. The arrival of an adult magpie at the nest was always accompanied by pandemonium among the chicks, as is entirely normal with most birds. The level of noisy confusion in the magpie nests we watched seemed especially pronounced, although this was probably due simply to the larger size of the species and the offspring's corresponding loudness. We could often hear the magpie nestlings long before we were close to the nest tree. This vocal onslaught must have been deafening for the adults, and grew ever more cacophonous as the chicks grew. Nonetheless, the parents labored away valiantly, returning again and again to the nest with their bills full of grubs, worms, caterpillars, moths, and skinks for the three or four ravenous craws in the nest. Watching from a safe distance through a powerful telescope, we were able to observe the continuous circuit of foraging on the ground until the bill was stuffed with squirming baby food, the direct flight back to the nest, a brief pause to unload into one of the noisy gapes, then a long swoop back to earth to start all over again. Male and female worked at almost the same rate, arriving at the nest in turns, on a punishing schedule that continued almost all day and lasted for about five weeks.

Breaks in this relentless routine of the adult magpies were infrequent but noticeable because they usually involved the birds flying out of sight from our particular vantage point. We simply assumed that they were somewhere within the territory that we did not have a direct view of. Subsequent investigation, however, revealed an unexpected discovery: these birds were partaking of the hospitality of some of the feeding stations nearby. We soon found that a large proportion of the pairs we were studying in the suburbs had the same arrangements: a conveniently positioned and well-provisioned larder. Not only that, this was a food source replenished daily, and because it was almost always within the boundaries of the territory, it was the exclusive domain of a local pair of magpies. And some pairs had at least three different stations to call on within their patch—although the people didn't realize that.

This was a major revelation for us and had both exciting and sobering consequences for our scientific investigations into the foraging ecology of suburban magpies. Our first important—and entirely unexpected—finding came through those careful observations of nestling feeding. As the chicks grew and their apparent hunger became audibly and visibly apparent even to us on the ground beneath, we assumed that the frantic and exhausted parents would take advantage of all that free food available nearby. Certainly the adults were visiting the feeding tables more frequently as the breeding season wore on. As explained previously, however, we found that almost all the food supplied to the nestling was natural invertebrate foods.[7] Even though they could so easily have ferried bill-loads of sausage and ground beef from the closest feeding tray, most of the baby food was the natural stuff. The human-provided foods were obviously being consumed by the adults, but even for them, it typically made up only a small proportion of their overall diet. This observation too, was unexpected: with so much easily accessible *human* food around, why didn't the birds simply forsake foraging for *natural* items altogether? These were important questions to our limited understanding of how the species was utilizing the available resources.

And our understanding was undoubtedly limited. Although we had made excellent progress in quantifying the productivity (in terms of measures such as worms per square meter) of different lawn types, sports fields, and farm pastures, and had gathered an enormous amount of data on foraging efficiency and chick provisioning,[8] the discovery that many of our birds were exploiting a considerable but unknown source of food entirely additional to their regular diet meant that our estimates of feeding resources were clearly incomplete and inaccurate. Attempting to understand the role of this new—"supplementary"—food source in the lives of our birds was now unavoidable. I had only been superficially aware of the many experimental studies into these issues, but now I really needed to find out more. Indeed, for reasons that will become progressively clearer, understanding the influence of bird feeding really starts with this type of research.

Food Is Fundamental

There are few requirements for life as undeniably essential or as influential as food in the lives of all animals. Food is the key to both survival and reproduction. Changes in its availability, quality, and the balance of its components can alter every aspect of their lives. The centrality of food can be seen in the extraordinarily detailed research being conducted into the nutrition of the animals entirely dependent on humans for all their needs. The food requirements of our pets and domesticated species are now understood in minute detail, allowing sophisticated and somewhat disturbing manipulation of such things as muscle development, milk composition, and fertility rates, all via their feed trays.

For wild animals, we know far less. This is unremarkable: there are a lot of species and they all have different diets. Unlike the economic motivation that propels commercial animal production, the reasons for attempting to learn more about the role of food for wildlife are more likely to be driven by scientific curiosity. Pragmatically, that usually means a lot less funding.

Nonetheless, researchers throughout the world have undertaken a vast number of studies on a bewildering array of species from tadpoles to polar bears, though a large effort has been focused on birds. The cumulative findings of these studies have fundamentally changed our knowledge of the roles that food plays, providing many insights into its function in the ecology and behavior of wild animals. As we will see, these apparently basic biological findings have been applied to many practical problems such as wildlife management and the conservation of threatened species.

Adding a Little Something

In one of the first comprehensive reviews of the research on supplementary feeding, the Canadian biologist Stan Boutin gets straight to the point: "No one would question that individuals and populations are ultimately limited by food supply."[9] The obvious difficulty is how to assess this food supply accurately. Even if we know what a species is supposed to eat, dealing with the daily and seasonal changes in availability, while accounting for competition and interactions between individuals, makes the task extremely complicated and logistically challenging. Most of the observational studies that attempt to account for overall food supply for a species readily acknowledge the limitations and unavoidable compromises.

The main alternative approach to straightforward observations is to intentionally manipulate the food available to an animal, usually by the provision of a suitable food that is extra to what would naturally be eaten. The key point is that this provisioning is additional, or "supplementary" to the background food resource. Some describe this as a feeding "subsidy," something supplied that, by definition, shifts the food supply to above that of natural levels. From a scientific perspective, such manipulations are often part of experiments designed to test hypotheses and provide potential explanations about the way food affects populations of wild animals. One of the most fruitful and enduring sources of such hypotheses was the research conducted by the famous British ornithologist David Lack, which he collated in two now classic books, *The Natural Regulation of Animal Populations* (1954) and *Ecological Adaptations for Breeding in Birds* (1968). Among many influential explanations for the cycles and dynamics seen in wild species was his suggestion that the fluctuations in the sizes of populations of many small animals were strongly reliant on the food supply available during the winter, the most taxing season. Lack's ideas have influenced generations of ornithologists, but the impact of *Adaptations* was especially pronounced. In 1969, the British Ecological Society brought together researchers to discuss the role of food for animal populations, with one outcome being an increasing awareness of the need for manipulation of food supplies in order to test ideas.[10] One of the powerful conclusions of this important meeting was that observations alone, no matter how detailed, cannot provide the explanatory power of a well-designed experiment.

Lack's ideas were especially linked to two of the most familiar, abundant, and well-studied species in Europe, the Great and Blue Tit. Along with the very similar chickadees of North America, tits (from "titmice," a general name for all members of the family Paridae) usually do not migrate and so are most vulnerable during the cold months in the temperate zone when winter foods are hard to find. Usefully, these small birds readily take to artificial nest boxes, which greatly assists researchers studying their reproduction. Both tits and chickadees also happily partake of human-supplied foods, another attribute of an ideal study species. Unsurprisingly, almost every aspect of the lives of these confident little birds has been carefully studied, including the role and influence of food. While a great array of species has been included in supplementary feeding experiments around the world, the prominence of tits and chickadees within this research field argues that we should pay them careful attention. That and the fact that these are probably the commonest visitors to bird feeders in the world. I think that we can legitimately nominate the "tit-adees" group as being eminently suitable representatives of the world's garden birds and spend some time reviewing what researchers have discovered.

Feeding Tits for Science

In and around the town of Ghent in Belgium, researchers have been studying the breeding of Great and Blue Tits using nest boxes first erected in 1959. Since those early days the number and diversity of locations with nest boxes have steadily increased until ten study sites containing over 800 nest boxes were being monitored. These sites spanned the general habitats typical of Western Europe: parklands in urban areas, suburbs with various deciduous tree species, and rural lands with mixtures of oak, beech, and pine. In the late 1980s, a group of Belgian scientists led by a young André Dhondt (remember the name) compiled almost two decades of detailed breeding information from these tits.[11] The researchers were especially interested in following up an observation made by other European ornithologists: that Great Tits living in city gardens tended to lay their first eggs of the season somewhat earlier than those nesting in nearby woods. The long-term information from Ghent allowed Dhondt and his colleagues to quantify these differences with some precision: across the sites, Great Tits in the urban parks started breeding an average of more than 10 days earlier than those in nearby rural sites. This was a remarkable difference. The effect was far less clear for Blue Tits, however, with the smaller species being later in the countryside but less pronounced in the other habitat types. The difference between the species and the habitats, the researchers speculated, was almost certainly due to the relative amounts of food being supplied by people; feeders were more often associated with urban gardens than were found outside the towns. Understanding whether this was the case would require experiments. As we will see, these were already under way.

At about the same time, ecologists working along the much drier and warmer Mediterranean coast of southern France were interested in the reasons influencing the start of breeding in the same two species, Great and Blue Tits.[12] Starting the breeding cycle early would appear to have obvious benefits to the parents: it allows more time to raise and fledge the nestlings, and possibly even start another brood. Earlier broods are also more likely to survive compared to later hatchlings, an outcome associated with having more time to find their main baby food of caterpillars, the most important source of natural protein for tits and many other small birds. Counting your chicks before the eggs are even laid is, however, risky. Females must be in a suitable physiological condition following winter before they can start the extremely taxing task of producing a clutch of eggs, and that would appear to be closely tied to the availability of suitable food resources leading up to the breeding period. To ensure that their females were in the best possible condition the French researchers decided to provide additional—supplementary—supplies from midwinter, long before the actual period in which the females started to develop their eggs. In addition, they altered the diet over time: in January, the feeding stations were provisioned with sunflowers and margarine, and a month later, dried insects were added. Then, about a month before the typical laying times, live insect food (mealworms, commonly fed and commercially available beetle larvae) was also supplied, to provide additional protein and energy for females during egg formation. All of these dietary additions stopped with the appearance of the first eggs so as not to influence other phases of the breeding cycle. Another (control) population within the same habitat but without the benefits of all that additional food was also carefully monitored.

By the end of this 2-year experiment, the local tits had consumed a total of 120 kilograms (265 pounds) of sunflowers, 8 kilograms (17 pounds) of margarine, and 19 kilograms (42 pounds) of mealworms. The main result of all that additional, high-quality food was an advance of about 6 days in the earliest eggs being laid by the Blue Tits in the study population. Not so for the Great Tits, however; there was a slight advance, but this was not statistically different from the dates of the unfed birds nearby. These results from the Great Tits are possibly even more surprising than those from the expected effect seen in the smaller species: despite all the additional foods (and both species really did consume most of what was supplied), the Great Tits did not respond in terms of when to lay. The reasons for such perhaps mixed findings were not understood with clarity but seem to indicate that food resources are only some of the cues that stimulate these birds to start breeding. Ornithologists have long argued that a complex suite of triggers may be involved, including changing day length (photoperiod), air temperature, female body condition, innate genetic predispositions, and even the availability of the artificial nest boxes. While these arguments have continued, the general importance of food in the overall breeding equation is clearly acknowledged as one of the main actors in a play with a sizeable cast.

Although similarly mixed results are standard fare for experiments involving birds in nature, it is still possible to claim that for the majority of supplementary feeding studies of titmice (and indeed, most birds),[13] some advance in laying date is a typical and expected outcome. This may suggest that birds breed earlier if they have extra food at critical phases of their lives (such as when they are engaged in egg production). However, it is also important to point out that almost all experimental studies report variable responses with some fed birds breeding much earlier than others. In addition, there is often considerable overlap in the timing of laying dates among the fed and unfed birds.[14] This rather typical finding may be due to the supplied food being insufficient in quality or possibly of little importance relative to the feeding opportunities available to the birds naturally. Don't forget that the foods being provisioned are, by definition, supplementary to the bird's regular diet. If the supplies of natural foods that the birds use are already high, the additional supplies may be simply less attractive—and less important in the broader scheme of things. This would indicate that there may be some important threshold value of food resources for many birds, which will result in a strong reaction (such as breeding much earlier) for those living below the level, but a much more muted reaction for those living above. In other words, if there is already enough food, birds may use different cues to begin breeding.[15]

Does Feeding Change the Timing of Breeding?

When to start breeding is among the most critical decisions any animal can make, but this is especially important for small birds like tits, which are only likely to have a few reproductive opportunities during their brief lives. As already mentioned, starting early has crucial benefits. For example, hatchlings born earlier in the season tend to grow more rapidly, and being in better condition than later-hatched chicks, they are more likely to survive and breed in the following year.[16] In contrast, chicks hatching later in the season are often less vigorous and already disadvantaged as they must compete with well-developed young that fledged before they did. The advantages of early hatching are, however, strongly reliant on their arrival in the nest from the egg coinciding with the peak in caterpillar supply, the primary source of nestling nutrition. This peak is typically measured in days, and so any situation that causes the birds to miss it, such as a period of poor weather conditions—or mistiming by the parents—can have catastrophic consequences. If birds respond to the manipulation of their food resources by breeding too many days earlier than normal, they could find themselves out of synch with the baby food supplies. This may be a legitimate concern for bird feeders: missing the peak by laying too early because the parents have responded to an enhanced food supply could potentially mean that feeding stations actually disadvantage the local birds. If so, this is an issue of fundamental importance.

Although attitudes—and organizational promotions—as to when to feed are certainly changing (as explored in Chapter 3), most wild bird feeding across the Northern Hemisphere has been typically limited to the winter months. Traditionally, this is probably based on the perception that birds may need some help to survive the tough times but that "it is better that they look after themselves" at other times. Because of the emphasis on investigating the role of food on breeding, a large majority of supplementary feeding experiments have supplied additional food during the early breeding season, and then assessed the effects on the birds' reproduction that followed. But most of these experiments do not replicate what happens when people feed birds. Does winter feeding actually influence the breeding activities of birds in the following breeding season? Do the effects of feeding carry over into the future?

Gillian Robb, under the guidance of Stuart Bearhop and Jon Blount, explicitly addressed this question in an important field study of Blue Tits in Northern Ireland.[17] This study was also significant in that it took a broad landscape-level approach in an attempt to more closely resemble the way feeders are distributed throughout the landscape. Ten separate woodland blocks already established with nest boxes were liberally supplied with peanuts (a commonly used bird food in this region) throughout the winter in one year but not the following year. Crucially, all feeding stopped at least 6 weeks before the first eggs were laid. This ensured that none of the female tits—fed or unfed—was influenced by food supply as they began to form their eggs, although their body condition was undoubtedly enhanced during the year of the study. The results were unexpectedly clear: feeding advanced the average laying date by 2.5 days compared to the nonfeeding sites, and, of particular significance, feeding resulted in more offspring being successfully fledged. Overall, feeding led to about one additional fledgling per nest, a remarkable outcome because this was not associated with any increase in the number of eggs layed or of hatchlings raised, both of which were unaffected. The extra fledglings resulted from their better survival compared to the unfed sites, yet this was obviously not related directly to additional food supply. The researchers put this effect down to the relatively higher body condition of the parents; healthier adults raise more kids. Winter feeding appeared, therefore, to lead to both earlier breeding and better survival of chicks. Given that we already know that feeding also improves the likelihood of survival of adults through the winter, these researchers also mention some potential implications of their work.[18] For example, places where winter feeding occurs may be producing both more new birds (offspring) and better surviving older birds (their parents) than places without feeders. This in turn may mean that the areas with reliable feeders are likely to support higher densities of longer-lived birds. This may be regarded as a welcome and possibly unexpected by-product of feeding in winter. For many species, being able to increase their reproductive output is certainly welcome news. But what about the still widespread perspective that we should be leaving the birds to themselves at other times?

As we have already discussed, the "winter-only" attitude is being challenged vigorously. Proponents of year-round feeding point to the apparent conservation benefits of studies such as that described above as why feeding should be continuous. Here we need to ask whether supplementary feeding experiments can shed light on this particular issue, given the proliferation of year-round feeding.

The best study to address this question was published under a title of commendable clarity: "Does food supplementation really enhance productivity of breeding birds?"[19] The question is fundamental to the conservation claims at the heart of continuous-feeding arguments and a test of the largely accepted assumption that more feeding leads to more chicks. Focusing on Great and Blue Tits using nest boxes, the experiment offered three feeding menus over a 3-year cycle but with the foods used swapping between sites so that the birds had a different diet each year. The key element of this experiment was 500-gram (17-ounce) blocks of protein-rich peanut cake continuously available from about a month before egg laying until up to 2 months after the end of fledging. Depending on the site and year, the birds received either peanut cake, peanut cake with mealworms added after the nestlings had hatched, or nothing (the control). While not strictly "year-round," the food was supplied during all phases of the breeding cycle, far longer than any other study on tits. Conducted by Tim Harrison, Jim Reynolds, and Graham Martin, the study took place in woodland sites in Worcestershire, central England.[20] Given the quantity and quality of the additional food being provided, the researchers were bold but justified in expecting big things: they anticipated that the fed birds would have earlier breeding, larger clutches, shorter incubation periods, enhanced hatching successes, and more chicks per nest. While these predictions tended to be a little optimistic (the effect of additional food has been quite mixed for clutch size, incubation period, hatching success, and brood size in a wide variety of species), very few studies had been as ambitious in the scale of provisioning. As a corollary, few studies were of such potential significance.

As would be reasonably expected by now, the date of laying was earlier and the length of incubation lower in the fed populations of Great and Blue Tits. Advanced laying dates are now a fairly common finding in experiments of this kind, although reduced incubation times are less so. What was not at all expected was the impact of prolonged supplementary feeding on the key components of reproduction: eggs, hatching success, and resulting brood size. Entirely contrary to the predictions, clutch size and brood size in both species were actually less for the fed birds compared to the unfed. Similarly, the proportion of eggs successfully hatching was lower in the fed Blue Tits, although there was no difference in the Great Tits. Thus, in relation to the question raised in the title of this study, the unavoidable answer was actually: "No, supplementation did *not* really enhance the productivity of the tits," although this conclusion relates only to this particular experiment. Nonetheless, the findings are reliable and sound and require careful consideration.

The results are also potentially alarming and sobering. They were certainly not anticipated by the experienced research team that conducted the experiment. Although one other supplementary feeding study also reported a decline in the number of eggs laid, it was a very different situation: a far larger, long-lived waterbird (the American Coot)—hardly an ideal comparison.[21] Where additional foods have affected clutch size, the result tended to be more eggs, but this is one component of breeding that appears to be difficult to alter. Most feeding studies found no change in clutch size, a finding that is even more pronounced for tits, with well over half of supplementary feeding experiments reporting no change in the number of eggs laid.[22]

What was far more significant in the Harrison study was the decline in hatching success in Blue Tits and brood size in both species associated with the provision of additional foods. These are unique findings among hundreds of supplementary feeding studies. This means that the extra peanut cake somehow led to a reduction in the proportion of eggs in the clutch producing live chicks compared to the unfed nests. The productivity of bird nests may be assessed at two critical intervals: first, at the time the hatchlings emerge from their eggs ("hatching success"), and finally, as young leave the nest as fledglings ("fledging success"). Between these two key milestones, plenty can go wrong. Starvation, predation, weather, and other factors will normally lead to a difference in the number of chicks that hatch and the number of fledglings that leave the nest. What is important for overall breeding productivity is the proportion of fledglings that survive, and that starts with the size of the brood. To have lower numbers of eggs and then fewer hatchlings is a major early blow, especially if this is due to the parents partaking of what should be of clear benefit to breeding.

It is important to remind ourselves that these are findings so far limited to one study in a single location over a limited time period. Nonetheless, the research was intentionally designed to assess the possible impact of more or less continuous feeding, as is increasingly the case in urban landscapes. It is striking, therefore, that some of the key findings of the Harrison study—earlier laying dates, smaller clutches, and reduced brood sizes—are remarkably similar to the breeding attributes of Great and Blue Tits living in typical urban environments in Britain. Compared to birds in largely rural landscapes, urban tits lay fewer eggs earlier and raise significantly smaller broods.[23] Furthermore, in a comprehensive review of the reproductive parameters of many bird species—including tits—living in urban areas, advanced laying and lower clutch sizes were found to be typical compared to birds living in nonurban areas,[24] a result attributed primarily to the greater availability of human-provided foods. This adds just that much more substance to the significance of the study we have been discussing, as well as raising some real concerns.

What about the Composition of the Food?

An obvious next step toward understanding the way supplementary foods may be influencing the breeding of birds is to see how characteristics of the food itself may play a role. At a rather basic level, the two fundamental components of foods are fats and proteins. Although both are essential, protein is especially important in preparing the female's body for the substantial demands of producing a clutch of eggs. A lack of protein—or, more specifically, particular molecules known as essential amino acids—during the early spring can, potentially, greatly limit the capacity of females to form their eggs, possibly resulting in fewer, smaller, or even poor quality eggs. Although these are common effects associated with seasons with low caterpillar numbers (a main source of natural protein), the same response has been attributed to the consumption of fat-rich but protein-poor foods such as peanut cake.[25] Indeed, the relative proportion of fat and protein in many commonly used bird feeder foods is decidedly fat biased; the percentage of fat and protein respectively for peanut cake is 70.5% and 17.1%, peanuts 44.5% and 28.7%, and black sunflower seeds 44.4% and 18.0%.[26] If birds are consuming more of these human-provided foods than natural insects, their bodies could be misreading the nutritional cues, with serious implications for breeding activities.

Fats are crucial as short-term energy sources and are almost certainly an important reason for the improved survival over winter for birds using supplementary foods. Most small songbirds, such as tits, however, are unable to store large, complex molecules such as fat in their bodies for any extended times. Instead, the fat-derived energy from all those suet balls and peanuts is utilized fairly promptly. And while winter feeding really does seem to enhance short-term survival, there is also some evidence that fatty diets generally can have longer-term health impacts. We know this from our own waistlines and heart disease statistics. What if this type of provisioning of birds for the tough times was actually detrimental for the birds we are trying to aid? Before we tackle that big question, a more immediate issue is whether supplying fats in winter affects things later in the year when breeding starts. After all, if fats aren't stored, perhaps there is little or no "carry over."[27]

These issues are, as usual, more complicated than they appear. For one thing, while macronutrients can't be stored for later use, a number of much simpler molecules known as micronutrients can be sequestered away and released when needed. This is especially important for reproduction in small birds because they are strongly reliant on the nutrients and energy immediately available as they start preparing for breeding. The most important of these micronutrients are calcium (needed for shells and bones), vitamin E, and a group called "carotenoids" (the source of most reds, yellows, and orange colors in animals and plants that can be stored in the fat reserves).[28] These molecules are antioxidants and play important roles in protecting the body from various physiological stresses associated with metabolism. The protective functions are increasingly valuable as the demands on the female's body build up during egg production. Vitamin E and carotenoids are also known to play critical roles as antioxidants in the developing embryos within the eggs, and the deposition of both micronutrients in the yolk has clear benefits for the subsequent body condition of the hatchlings.

These somewhat technical details are needed as background to discussing the next crucial supplementary feeding experiment. Undertaken by Kate Plummer and colleagues, this particular study is of fundamental importance because it explored the potential carry-over influences of winter feeding on breeding in Blue Tits, with careful attention to the way that the macronutrients (fat) and micronutrients (vitamin E and other carotenoids) affected egg production.[29] In other words, the experiment replicated typical winter feeding of small birds throughout the Northern Hemisphere.[30] Again, the strength of this research stems from the clever but simple study design employed. The researchers set up three nest box sites in deciduous woodland in the beautiful countryside of Cornwall in the far southwest of England. Three different menus of supplementary foods were used: fat alone (handmade balls of vegetable fat); fat plus vitamin E (added to the ball at the same concentration as that found in peanuts); and, of course, no additional treats at all (the control) as a comparison. These courses were supplied for the winter period only and were stopped a month before the start of the breeding season, well before the birds had even started to think about breeding. Because the key goal was to assess whether the effects of the different winter-feeding regimes carried over into the following seasons, the diet offered to the birds in each study site was swapped each winter over the 3-year study. If there were clear outcomes, these should show up in the eggs laid in each of the sites according to the diet.

Following on from our discussion of the ability of birds to store nutrients needed for breeding, we would probably expect that the provision of fat alone—so early in the cycle—would not affect egg production, while the addition of vitamin E would probably be useful. The findings of the experiment were, again, unexpected. First, contrary to other similar studies, there was no clear change in laying date. Nor was the number of eggs produced or their relative size affected by any of the diets. What was thoroughly unexpected, however, was that both the amount of yolk per egg and the level of stored carotenoid were significantly lower for the birds consuming the fat-only diet. These components of the eggs are critically important to the growth and development of the embryo and the health of the chick. The yolk contains all the ingredients needed for the construction and maintenance of the growing offspring within the egg. For this reason, the body condition of the mother bird *at the time the eggs are formed* appears to be of fundamental importance to the subsequent well-being of her young. Therefore, to find that partaking of a fat-heavy diet in winter appears to result in female tits producing impaired yolks—and potentially leading to lower-quality chicks—is of considerable concern.

But what happened when vitamin E was added to the fat balls? Although theoretically some benefit of the addition of this antioxidant was expected, the outcome observed was still striking: vitamin E appeared to cancel out the effect of the fat, with yolk size in the groups of birds receiving this diet being no different from that of birds eating a natural diet.[31] How might this come about? Although the actual biochemical interactions involved were not studied, enough is known about fat and antioxidants to be able to offer plausible explanations as to how this might work.

As explained earlier, small birds cannot store macronutrients such as fat for long periods: it is a useful form of energy that enhances the physiological challenges of surviving the cold months. Nonetheless, access to readily available, high-fat foods in winter may mean the birds are less likely to be obtaining a more diverse natural diet. This may also mean they have fewer antioxidants, at a time when they are internalizing plenty of polyunsaturated fats, just the sort of harmful molecules the vitamin E and other carotenoids can neutralize. Although a fatty diet in winter would not have directly influenced the formation of yolk much later, the birds' bodies appeared to be still dealing with the oxidative stress, which affected their ability to form egg yolks. By adding an antioxidant, this stress seemed to be reduced.

Yes, we certainly need to be careful in extrapolating generalization from a single experiment, however elegant. These are results specific to Blue Tits in Cornwall, and the study has yet to be replicated anywhere else or with other species. Equally, however, it would be sensible to learn from such carefully planned and relevant research that not only looked at a popular feeding practice (providing fat in winter) but may also offer a solution (adding an antioxidant). For tits, at least, this may be as simple as throwing some peanuts in with the suet (although, of course, this still needs verification).

Feeding and Survival

Finally, we return to the primary question of the relationship between food and survival in these tiny birds, the key to population regulation as suggested by David Lack, and perhaps the fundamental issue associated with supplementary feeding experiments. Indeed, virtually all the features found to be influenced by the addition of food may be evident in the number of birds remaining in the population in the year following the experiment. If, because of the availability of additional food, more eggs are being laid by healthier females, and if the resulting hatchlings are slightly more likely to make it through to fledgling, and if these are more likely to survive the winter, then it follows that there should be more birds than before the food was provided. Yes, theoretically, but there is plenty that can go wrong too.

As portrayed by Lack, the harsh realities of winter in the Northern Hemisphere's temperate zone provide the ultimate test of the population that survives to breed in the following spring. As conditions become increasingly tough and foods scarce, competition for what little is available is inevitable; food supplies will pretty much determine survival. Numerous supplementary feeding studies have assessed this directly, providing additional food and monitoring the changes in bird numbers before and after. In an early study conducted near Lund, Sweden, Great Tits at several sites were either provisioned with lots of sunflower seeds (dispensed in large hoppers) or had to make do with completely natural resources.[32] The number of pairs were counted in each site over the two following winters and showed dramatic yet dissimilar results. During the first winter of the study, which happened to be particularly severe, populations of tits without additional foods decreased by at least 10%, whereas the number of birds at the fed sites increased by between 20% and 60%, depending on the site. Clearly, supplementary foods were greatly enhancing the ability of the birds to survive the winter. It also greatly improved the chances of young birds hatched the previous spring to remain through the winter and go on to breed. But the story gets more interesting. In the following year, all populations, whether fed or not, increased dramatically, with numbers of tits being between 35% and 94% higher than the previous winter. This spectacular result clearly had little to do with all the seed laboriously provided by the researchers. Rather, natural events that year entirely overwhelmed the influence of careful experimental design. The winter months in southern Sweden that year were distinctly mild with considerably less snow, enabling the birds much more opportunity to forage naturally. But far more important, that year the local beech mast crop was large, providing an abundance of easily obtained natural food. In such circumstances, to paraphrase the author of this study: "In a good mast year, there was no effect of supplementary food."[33]

By this stage the attentive reader will have noticed, possibly with some affront, that this discussion of feeding experiments, ostensibly covering the titmice group (the parids), has been entirely preoccupied with European tit species. But what of their close North American relatives, the chickadees, and especially the ubiquitous Black-capped Chickadee? Unfortunately, far less supplementary feeding research has been conducted on these extremely popular little birds, a somewhat surprising situation given that they are almost certainly the most abundant species using feeders throughout the United States and Canada. Black-capped Chickadees are especially conspicuous during winter in the more northern parts of their distribution as they are among the smallest birds to remain during winter and consequently flock to feeders in large numbers. The potential value of human-supplied foods during winter for chickadees has long been recognized, but was most famously studied by Susan Smith in the Massachusetts woods during the 1960s. Professor Smith would continue to investigate every aspect of the behavior and ecology of these birds for the next 30 years,[34] but it was her first study of overwinter survival that is of particular importance to this discussion.[35]

To estimate the impact of winter feeding on birds, it is important to compare the number of birds before and after. For most studies, this involves attempting to count birds visually using standard methods that ensure that the place, duration, and searching effort are the same each time. For anyone who has watched flocks of chickadees (or most small, active birds) swarming around feeders, then sweeping into the nearby foliage before trickling back in dribs and drabs, making a reliable "count" is, shall we say, challenging. One of the main problems is that we don't know how often the same birds are being recounted. This inevitably makes comparisons rather sketchy, undermining our ability to make sound claims. The significance of Susan Smith's remarkable research is that her estimates of numbers before and after winter were unusually robust. She was able to identify individuals among the clouds of otherwise anonymous chickadees because most of the birds had been marked with individually colored leg bands (or "rings"). Although color-banding is thoroughly standard practice for much bird research today, Susan Smith's ability to recognize specific birds was fairly revolutionary at the time.

Far more prosaically, however, individually marking her chickadees allowed Susan Smith to show that survival over the winter was largely determined by the reliability of the local food supply. Being able to identify known birds confidently at regular intervals throughout the winter and again in the following year allowed careful accounting of the ongoing presence—or sudden disappearance—of each bird. But while Smith could claim that providing additional foods aided chickadee survival through the winter, unfortunately the lack of a "control" site where food was not supplied meant that little could be stated about the relative influence of supplementary foods.

Smith's studies provided a solid foundation for the next phase of American chickadee supplementary feeding studies. During the 1980s, Margaret Clark Brittingham and Stanley Temple from the University of Wisconsin explicitly addressed the limitations of the earlier work by thoughtful yet straightforward experimental designs.[36] In the first of several studies of great importance to our understanding of wild bird feeding, Brittingham and Temple color-banded 576 Black-capped Chickadees in five well-separated study sites in rural Wisconsin.[37] Although chickadees are known to be fairly sedentary, remaining close to home year round, it was still remarkable to learn that only two of the almost 600 marked birds were detected more than 2 kilometers from where they were first captured.

Using techniques now familiar to us, chickadees at three of the sites were supplied with a feeder regularly filled with sunflower seeds from October to April, while those at two other sites were offered nothing. This continued for 3 consecutive years except for the crucial variation of stopping the food supply at one feeding site (although, tantalizingly, the empty feeder remained in place) and starting feeding at a site previously without a feeder. As a powerful attempt to understand the impact that the provisioning—and withdrawal—of additional foods may have on the survival of tiny nonmobile birds, this should do it. Their persistence as winter progressed was determined by thorough weekly searches of each site for marked birds and regular captures using mist nets. The detailed information collected allowed monthly survival rates to be calculated as well as year-to-year survival. In addition, astonishingly precise data (to within one-hundredth of a gram) on the body masses of the birds were obtained at the time of capture. Though the procedure is not simple, the painstaking efforts required to handle such a small yet spirited creature were well worth it. Perhaps surprisingly, small birds such a chickadees gain weight over the course of a day as they forage and then lose the weight as they use up their meager stores of fat during the night. The difference in mass at dawn may be a much as 10% less than the night before. For birds weighing only about 12 grams (0.4 ounces), these differences may mean the difference between life and death on a long, freezing night.

It was in the body mass data that the first significant result was noted. At the start of the experiment, males and females were on average slightly lighter (0.17 grams, or 0.006 ounces) than later in the winter, but birds with access to supplementary food were just a little (0.13 grams or 0.005 ounces) heavier than the unfed birds. The differences may seem small, but they appeared to translate into significant survival statistics. Chickadees supplied with additional food had, on average, a 95% chance of surviving through to the next month and a 70% chance of making it all the way through the winter. For the unfed birds, this compares reasonably well, with a monthly survival probability of 87%, but contrasts starkly with an overwinter survival of only 37%. The benefits of having access to supplementary food is even more obvious when monthly survival for fed and unfed chickadees is compared during severe winter conditions (periods of more than five days per month with minimum temperatures less than –18 °C [–0.4 °F]). Without the supplements, about a third of the chickadees disappeared (presumably died) during these cold snaps, while only 7% of the fed birds succumbed.

While these stark numbers may convey something of the scientific significance of the experiment, it is the author's observations of the behavior of the birds that we find especially compelling. Recall that chickadees are tiny, continually active birds, quite unable to store sufficient fat to last more than a full day without foraging. Yet Brittingham and Temple observed the birds during a period of prolonged extreme cold when the maximum temperature did not rise above –18 °C (–0.4 °F) and the minimum reached –29 °C (–20.2 °F) for over five consecutive days. Although writing in the objective prose required by scientific publications, the authors cannot disguise the reality of what they observed:

> During such periods of extreme cold, the behavior of the chickadees on control sites [the nonfed birds] changed. We could not find most individuals, and the few we located were extremely lethargic. They sat motionless, with feathers fluffed out, facing the sun.[38]

The researchers speculated that this marked inactivity, so unlike the usual demeanor of the species, was their only remaining physical response to the severe cold. To move would expend energy they did not have; by remaining motionless they were conserving the little reserves they retained. Whether this was a strategy that worked would be measured by the brutal calculus of each bird's fat reserves, the duration of the extreme conditions, and the availability of suitable nutrition once the weather improved. In the frigid midwinter woods of Wisconsin, the odds would not seem favorable.

Unless there was a feeder nearby. With access to an unending supply of sunflower seeds, the behavior of the chickadees lucky enough to live near such a bonanza was shockingly different from that of their hard-pressed and unfed colleagues. Despite the severity of the prolonged bitter conditions, "[fed] birds continued to use the feeders normally."[39] The contrast could hardly be clearer: the additional food made a very real life and death difference to these birds.

These are strong findings, but are not dissimilar to other winter-feeding studies on similar species. For example, overwinter survival of two species we have not yet mentioned, Crested Tits and Willow Tits, studied in Sweden was almost double that of birds without access to supplementary foods.[40] It should be noted, however, that some studies that have investigated the effect of additional foods on winter survival have not reported such clear findings.[41] Indeed, some experiments report no influence on survival or on any of a long list of the other parameters mentioned above. These studies may be just as important as those reporting dramatic effects and point to the realities of scientific field studies: the results rarely play out as simply as expected. This is a theme we will return to at the close of this chapter.

Changes in Behavior

To conclude this selective review of supplementary feeding experiments on tits and chickadees, we turn to the possible influences on perhaps the most malleable and immediate features of their lives, their behavior. It is one thing to measure physical attributes such as egg size and body mass; it's a very different matter to ask questions about what the birds are actually doing in response to the additional food.

Given the importance of food to the day-to-day survival of these small birds, particularly during tough conditions, it would hardly be surprising to find feeding can change their social interactions. Although an obviously social group, spending almost all their time in loose groups, tits and chickadees tend to forage alone when seeking their natural diet of insects and seeds. They may be close to the rest of their flock, but the type of food they glean means that it is best obtained without the interference of other birds. During periods of natural food abundance, when there is little competition between individuals, the birds are fairly tolerant of one another's presence. As conditions become more difficult, as occurs with the start of cooler weather, relationships often become less amicable. Wilson looked at such interactions among her Black-capped Chickadees during the harsh winters in Maine when feeders were experimentally introduced.[42] In natural situations, the onset of cold conditions led to increasingly more assertive behaviors by individuals with smaller foraging areas being defended against other chickadees. These patches were established throughout the site, with each bird actively patrolling what was now a small but private "foraging territory." Trespassing was strictly policed, each bird the master of its own modest domain.

The introduction of feeders abruptly changed all that. With the sudden arrival of a super-abundant food bonanza at just a few localized spots, the former system of territorial enforcement broke down entirely. For the birds living close to the feeders, whatever perceptions of outrageous good fortune they may have had soon dissolved in the reality of the inevitable influx of all and sundry to the feeder. Although many of the birds initially attempted to maintain their boundaries, the constant crossing by birds drawn to the feeders soon led to territoriality being given up completely. For these small birds, there was simply no way that an individual could successfully defend "their" feeder. In other situations (discussed elsewhere in this book), usually where the species are much larger and more capable of effective defense, the introduction of feeders can lead to a serious escalation of territorial behavior.

Another influence of supplementary foods on foraging behavior relates to joining feeding flocks made up of different species. While it is generally rare for most birds to deliberately seek the company of other species, the formation of groups of similarly sized birds of a variety of species has been reported from around the world. These flocks are temporary and are usually made up of groups of insectivorous species that occur together in the same woodland or forest habitat. Although there is much we don't know about these strange amalgamations, most ornithologists believe that birds join them as a way of improving their ability to find food in times of scarcity. This idea has been tested by researchers working on tits in the UK,[43] as well as yet another parid species, the Varied Tit, in Japan.[44] In both cases, providing additional foods led to a significant decline in the tendency of birds to join these mixed-species flocks. Interestingly, birds partaking of the supplementary foods were more likely to forage alone when away from the feeder.

One of the most conspicuous behaviors of songbirds such as tits and chickadees is their vigorous singing, particularly during the early stages of the breeding season. As pleasant as these sounds may be to our ears, the functions of these impassioned displays are complex and deadly serious for the (usually) male participants. While female songbirds also vocalize, it is almost always the males that produce most of the early morning noise. While we can never be entirely certain of what these vocalizations mean to the birds themselves, countless studies have established that birds use song for a range of purposes but especially to improve their chances of reproduction. Songs can be about territorial defense, the proclamation of identity and occupation of a space, as well as an invitation to mate. Such sweet melodies (to us) may be adamantly "Keep away" (intended for the ears of other males) as well as a vocal résumé of prowess to be assessed by potential female partners. Yes, both of these general functions can operate at the same time, although once the male singer has successfully attracted a mate, the song's territorial function becomes his key modus operandi.

The introduction of a feeder into this scenario is likely to have a number of consequences. As we have already described, birds near a rich food source may become far more assertive in their territorial defense as they attempt to repel the inevitable increase in intruders. In plenty of studies, males with territories near feeders physically attacked trespassers and sang more often.[45] Fighting and singing are both activities requiring considerable energetic resources and normally take up important time that could be otherwise spent looking for food. With a regular supply close by, males near feeders have the fuel to sing more than those forced to forage naturally. This food-enhanced singing may have real implications because females of both Blue Tits and Black-capped Chickadees have been shown to prefer males that sing earlier in the day and who sing more.[46] Indeed, this attraction continues even after pairs have been formed, with Blue Tit offspring not fathered by a female's mate most likely to be the apparently sexy singer up the road.[47] Unsurprisingly then, supplementary feeding has resulted in earlier dawn singing and greater song output in a number of species (including Common Blackbirds in the UK[48] and Silvereyes in New Zealand[49]).

Working within the suburbs of Oslo in Norway, Katja Saggese and colleagues were interested in the possible changes to the vocal behavior of Great Tits associated with supplementary feeding and the possible ongoing long-term effects of feeding.[50] These researchers provided ad libitum sunflower seeds and fat balls to their birds and recorded the songs of males either side of dawn, noting the exact time of all relevant events. Great Tits typically start to sing vigorously at least six weeks before egg laying, well before dawn. Would the fed birds start even earlier? Or would the extra food lead to more singing as predicted by previous studies? To the researchers' considerable surprise, male tits with access to additional foods actually started to sing later than unfed birds. And they did not make up for sleeping in by singing more; their song output was no different from their hungrier colleagues. Again, these are important findings that deserve some consideration. The people who conducted this study were refreshingly candid about their results being unexpected, but they did offer some plausible explanations.

Waking well before dawn, by definition, means moving about in virtual darkness, and it has been suggested that attempting to find their insect foods at this time would be very difficult.[51] Singing, with the various benefits mentioned above, may actually be a more sensible activity. Birds with access to feeders, on the other hand, would probably be able to feed even in poor light and may be feeding instead of singing. In addition, what of the apparent reproductive advantages of early singing mentioned previously? To ignore aspects such as attracting the attentions of other females suggests that males may be weighing up their options. For example, having access to such a rich and reliable food resource may itself be a powerful inducement. Alternatively, females paired with feeder-owning males may be less inclined to wander. In reality, these are all simply speculations requiring further careful study. Yet again, this nicely executed feeding experiment has delivered intriguing findings that hint at the complexity of the situation. Again we have to conclude that feeding changes things in unexpected ways.

The bulk of information presented in this chapter has focused intentionally on supplementary feeding experiments conducted on tits and chickadees. This somewhat justifies our focus on this group, but there is obviously more to the field than these selective summaries suggest. For example, the significance of supplementary feeding experiments to wildlife management and conservation are such that we devote another chapter to these major themes. To conclude this exploration of supplementary feeding research on garden birds, however, we turn to some key studies of species other than titmice.

Scrutinizing Scrub Jays

One of the more unusual species visiting bird feeders in North America is the Florida Scrub Jay. Although a member of the superabundant and familiar corvid group—the crows, ravens, and (true) magpies—the Florida Scrub Jay is in serious trouble. At home in the sandy, dry scrublands of central Florida, this exceptionally intelligent and highly social species faces habitat losses and degradation due to expanding housing developments, hotels, golf courses, and citrus plantations. As a result, the species has been steadily declining over the past few decades and currently exists in a small number of isolated populations—always a situation that raises the chance of local extinction. Plenty of other species face similar predicaments, and our understanding of their situation is limited. By contrast, we know a lot about the behavior and ecology of Florida Scrub Jays, and our knowledge provides an unusually detailed background for developing sound conservation plans. A critical component of this information is based on numerous supplementary feeding studies.

I described Florida Scrub Jays as "unusual" feeder visitors not just because they are so obviously different from the more typical garden species and their dire conservation status but also because of their remarkable social lives. There cannot be many places in the world where people living in the suburbs play host to a threatened species that also lives in permanent extended families, with all the associated dramas. It was the intriguing communal breeding arrangement of Florida Scrub Jays that led to the pioneering field studies by the late Glen Woolfenden and John Fitzpatrick (now at the Cornell Lab of Ornithology) conducted at the Archbold Biological Station in central Florida. These now famed studies, which commenced in the 1970s, described the remarkably complex social relationships based around the presence and active involvement of individuals other than the breeding pair.[52] So-called helpers at the nest, these birds were discovered to be the young from previous years who remained with their parents to assist in the raising of the next batch of young. Woolfenden and Fitzpatrick's influential research was instrumental in kicking off worldwide interest in the phenomenon now known as "cooperative breeding."[53]

Stephan Schoech, Reed Bowman, and Jim Seymour are three researchers engaged in long-term studies of Florida Scrub Jays who have taken a particular interest in the influence of food and nutrition in the lives of these birds. In an initial study in the 1990s, Schoech supplied dog food, peanuts, and mealworms (a diet fairly similar to that available to the Scrub Jays in nearby suburban developments in Highlands County) to some groups and not to others, all birds living near the Archbold Biological Station.[54] After decades of continuous interaction with researchers, the birds were only too willing to participate in the study. As Schoech explained: "Group members quickly learned where the feeding station was and if they were not already waiting for me, rapidly responded to my whistle."[55] The results were spectacular: breeding was advanced by an average of 16 days. (Recall that we were excited about differences of just a few days in fed tits.) Obviously, the provision of additional foods is of primary importance to decisions of when to start breeding in this species. Interestingly, the study also tested one of the key hypotheses about the causes of cooperative breeding: helpers don't breed because there is insufficient food. However, although two of the fed female helpers did forsake the life of an au pair, found a partner, and went on to breed themselves, the others (thirteen of fifteen) did not. It would appear that among Scrub Jays, there is more to such a fundamental change in behavior than simply food supply.

These findings have obvious implications for Scrub Jays with access to the well-filled suburban feeders, increasingly available as suburban developments expand in central Florida. Generally, these birds do not live in suburban areas but visit periodically from their territories in the scrublands nearby. This proximity to a year-round supply of foods such as pet food, peanuts, and food waste led to about a third of the diet of breeding female Scrub Jays being provided by people.[56] As expected, these birds started to breed earlier compared to their scrubland relatives.[57] Presumably, this was due to the relative reliability and higher fat content of the human-provided foods, but teasing this apart requires more control over the variable; that is, a proper supplementary feeding experiment.

Jim Reynolds from the University of Birmingham in the UK (whom we met earlier) is a long-term collaborator with extensive experience working with the Scrub Jays. An experiment he conducted with Stephan Schoech and Reed Bowman was designed to assess the influence of different fat and protein levels within the foods provided.[58] Of course, these issues had been previously studied by other researchers, using the usual commercial seeds as supplements and assuming standard nutrition content. What Reynolds and his colleagues wanted was far greater control of the actual levels of the fats, proteins, and carbohydrates present in the foods provided. To do so, they had their own made, arranging for a commercial operator to produce vast amounts of little cylindrical pellets to a very specific recipe. These were of two "flavors": high fat and high protein, or high fat and low protein. Both types had, however, identical energy content. This was supplementary feeding science on a new level altogether.

Clever, yes, but only if the Scrub Jays agreed that the odd-looking, gray-brown pellets appearing in their feeders were edible. After all, they looked nothing like the dog food, peanuts, or mealworms they were used to. This particularly adventurous species had no hesitation, however; they loved the pellets and the experiment was off to a good start.

As usual, Scrub Jay territories were randomly assigned to either high fat, low fat, or no fat (that is, the "control" group, who were offered welcoming but empty feeders). Supplementary feeding started in midwinter and continued until just before the start of laying. In the second year, everything was again randomized and no group received the same diet two years in a row. Again, the results were remarkable and, because of the sophistication of the experimental design, the particular dietary influences could be teased apart. First, as we would now expect, the start of breeding was considerably earlier for fed birds compared to the unfed, although again the extent was extreme: on average almost two weeks earlier. What was unexpected, and to my knowledge has not been shown in any other supplementary feeding study, were similarly strong effects on three other fundamental breeding characteristics: clutch size, egg mass, and egg composition. Unfed Scrub Jays produced an average of 2.8 eggs while those on the low-protein diet laid 3.8 eggs and those on high protein laid 3.5 eggs, although this difference only occurred during the first year. This is especially remarkable given that the normal clutch size for this species is only three.

But simply laying more eggs does not necessarily translate into more chicks: the survival of hatchlings is closely associated with both the size of the egg and the way it is provisioned for subsequent embryo development by the female. In most birds, egg size declines steadily with each additional egg produced, with the last laid being the least likely to survive. While a steady decline in the mass of eggs as each was laid did occur in each of the three dietary treatments, Scrub Jays on the high-protein diet were able to maintain the mass so that the third egg produced was actually much heavier than the equivalent egg laid by each of the unfed females.

Studies such as this Scrub Jay experiment indicate an increasing interest by researchers in some of the finer details of the relationship between food and a wide range of aspects of animals lives. Carefully designed and well-conducted experiments on wild species have been essential to uncovering the links in this complex chain of influence and consequence. Additional food almost always changes something. As we have discovered here, supplementary feeding experiments have also revealed unexpected and unwelcome findings. Some of these results may well alter the way we think about and practice our own wild bird feeding.

Maggie's Meat

As a final example of the way that our feeding can affect garden birds we return to the unusual scene at a typical feeding station in an Australian back yard. Instead of sunflowers, peanuts, and suet there is often salami, ground beef, and cheese. And that's because rather than birds like tits, chickadees, or even Florida Scrub Jays, the most likely visitors are all large, assertive, and often—strangely enough—black and white, with a distinct taste for meat (though just about anything vaguely edible will do). These pied picnickers may include (Pied and Gray) Butcherbirds and (Pied) Currawongs, as well as (the non-pied) Laughing Kookaburras. But far more probable are the Australian Magpies introduced at the beginning of the chapter. Not only are these big, bold birds the commonest species at Australia feeders, they are also the most welcome. More people try to attract "Maggies" than any other species: hence all that meat.

As demonstrated by numerous examples described already, researchers are often intrigued, if not concerned, by the physiological impacts of anthropogenic foods on wild birds. Such was the case with the overly meaty diet of Australian Magpies being fed throughout the suburbs of the country. In an important experiment conducted by Go Ishigame and colleagues from the University of Queensland, three types of common foods—minced meat (ground beef), pressed dog food ("dog sausage"), and shredded cheese—were supplied to groups of free-ranging magpies.[59] These were typical foods available at feeders used by local magpies. Experimental provisioning occurred over a series of several month-long sessions, during which each group received one food type or none. Rather than the usual breeding parameters we are now familiar with, the focus of this experiment was on blood chemistry. Critics of wild bird feeding have suggested that some of the foods being consumed by urban birds may be harmful to their health. Certainly, the foods offered—and accepted—by magpies could be included in this category. To assess this, at the end of each monthly feeding session, Ishigame and company caught the birds and obtained tiny blood samples, which were tested for several parameters.

Interestingly, both cholesterol and uric acids levels were found to have increased significantly but only for the dog sausage. Both parameters are well known as indicators of negative health conditions. The result is also noteworthy because the birds consumed similar amounts (about 40 grams [1.4 ounces] per day) of each food type, yet the mince and the cheese had no discernible effect. This finding may be associated with the relative level of commercial processing involved, with both the meat and cheese being considerably simpler foods compared to the dog food. Indeed, the fact that this item is manufactured and marketed as a balanced pet food—albeit obviously for dogs—but produced a marked spike in cholesterol levels in birds may be of some concern. And while plenty of magpies are known to partake of such pet foods—both as supplied by feeders as well as by sneaking bits from pet bowls—there are also vast numbers of birds being provisioned daily with processed meats such as salami, German sausage, and other cholesterol and fat-rich fare. And we already know what happens when these foods are eaten to excess in human diets.

Food for Thought

That's probably enough examples of supplementary feeding experiments for now. Those featured here are only a fraction of the many such studies that have been published, but they represent a fair selection of those that relate directly to urban birds. As we are obviously interested in understanding the influences of providing additional food to wild garden birds, those dealing with other bird groups or topics associated with conservation and wildlife management were not included here. Furthermore, the studies featured have tended to be those with the more marked results. That is hardly unusual as it's always easier to write about—and more interesting to read—the research where the findings were spectacular or unexpected. Nonetheless, it is also important to be aware that plenty of well-designed and carefully conducted experiments did not find an effect. I suspect that many such studies simply do not get published, leading to a definite bias toward positive and significant findings in the available literature. Some, however, have made it through the rigors of the peer-review process and these also need to be acknowledged.

Although there are several excellent reviews of supplementary feeding experiments available (see Boutin's 1990[60] review for many examples), a particularly readable summary of relevance to this discussion is that of Gillian Robb and colleagues (whose study of winter feeding of tits in Ireland we described earlier).[61] In their article, titled "Food for thought," the authors reviewed as many supplementary feeding studies on birds as they could find and determined whether the addition of food resulted in either a positive or negative or no response on each of seven commonly measured breeding parameters. Although these researchers looked at a number of different bird groups, we will concentrate on their "small passerines" category, as the most representative of the species that come to feeders.

The breeding parameters featured in "Food for thought" were as follows: laying date, clutch size, egg size or quality, incubation time, hatching success, chick growth rate, and fledging success. Considering the direction (positive, negative, or none) of the response of all these features together, two conclusions are immediately clear. First, negative responses to supplementary feeding were very rare; only a single study, reporting a later laying date with feeding, was cited. Second, the most common response, by far, to supplementary feeding among small passerines was none. Only for laying date and fledging success did positive responses outnumber no response. While this was most pronounced for laying date effects, with eighteen out of twenty-eight studies reporting earlier egg production due to feeding, almost a third of studies found no effect. Bear in mind that an advance in the timing of breeding is certainly the most publicized effect associated with the provision of additional food. These studies suggest that this commonly reported outcome is far from ubiquitous. Indeed, when all the bird groups included in the review are considered, thirty-four studies reported earlier breeding while twenty-four—that's 40%—found no change at all. For the only other breeding parameter where a majority of studies found a positive response—fledging success—the ratio was even closer: ten were positive compared to nine with no response. Food for thought indeed.

Good News and Otherwise

We have traversed a lot of territory in this chapter, providing perhaps too much detail to remember. The main message to be gained from these experiments is that feeding garden birds often really does change things. For example, feeding does assist survival through the winter and especially during periods of prolonged severe conditions and lean years when natural foods may be scarce. Winter feeding also leads to better survival of chicks, due in part to the elevated body condition of the breeding females. Providing additional food may therefore lead to the presence of more birds locally, through the attraction of visitors to an abundant foraging resource as well as through the enhanced breeding of the residents. All of this is surely good news. However, these changes may also include unwelcome outcomes we did not expect.

One of the roles of scientific rigor is to ensure that we see beyond what we hope is happening or what we might logically expect. Certainly, some of the researchers conducting these supplementary feeding studies were surprised—and sometimes alarmed—at their own findings. For example, to discover that a study replicating typical continuous ("year-round") feeding practices actually led to fewer offspring is of considerable concern. This is because the experimental results appear to mirror what is now widely appreciated more generally: that suburban birds tend to lay fewer eggs and have smaller broods than rural populations of the same species. Feeding is directly implicated. Similarly, studies mimicking the fat-heavy diets of many typical feeders found that this can impair yolk and embryo quality, with real risks to the survival chances of the hatchlings. Again, a worrying result, although this important study may also have come up with a possible solution (the addition of an antioxidant such as vitamin E).

This is a suitably sobering note on which to conclude our discussion of the science of supplementary feeding. As we have seen repeatedly in the examples described here, a proper scientific approach requires a genuine willingness by researchers to be critical, clever, and open-minded. When conducted with appropriate rigor and attention to detail, careful experimentation may be essential for progressing ideas beyond what is possible through observations alone. This is certainly the case for anyone seeking to understand the apparently obvious, actually complex interaction between birds and the foods we provide for them.

So, in conclusion: feeding almost always changes things although sometimes in ways that are different from what we were expecting.

6

Tainted Table?

Can Feeding Make Birds Sick?

I have met plenty of keen bird watchers who also feed, but none come close to Frank Wilston for infectious enthusiasm and scientific dedication. I visited his home in a small town outside Washington, DC, late in spring and was overwhelmed by his hospitality, master naturalist's knowledge, and wicked sense of humor. Having just celebrated his 85th birthday (and shared his cake with some excited Tufted Titmice, perched on a steady hand), Frank admits that he doesn't get out birding in the nearby woods as much as he would like these days. A self-confessed "avid birder" (he says his friends are more likely to say "darned fanatic") from his earliest days, Frank now channels his considerable energy into ensuring that the birds come to him. This has meant a fundamental redesign of his large yard, with every feature—the position of the garden beds, the selection of each shrub and tree, the numerous "no-mow" sections full of grasses and weeds—strategically positioned with viewing birds in mind. Fundamental to this aim are the feeders. As Frank guides me through his remarkable garden, I try to keep count: there are lots of tube hangers of all sizes, including fine-mesh tubes for the tiny nyger seeds, as well as large-mesh peanut balls. There are also several simple platforms, a large and fancy transparent hopper feeder, several "thistle" socks, and at least a couple of little feeders stuck to the windows. "I think that comes to fifteen feeders, Frank," I venture. "That means buying an awful lot of seed." "You missed the two 'woodpecker feeders' in those trees yonder," states Frank matter-of-factly, pointing to a stand of Scots pine. "And yes, that *is* a bunch of birdseed. It costs me a packet, but I'm happy to spend my hard-earned money on these delights. Better that than let my lousy relatives get it!"

It's late afternoon when we settle into a couple of chairs on the high-set veranda overlooking Frank's marvelous garden, home-brewed beers in hand. As I glance around the yard, the genius of Frank's attention to the placement of his feeders is obvious: from this single vantage point I can see eight, at different heights and settings, providing a lot of foraging—and viewing—opportunities. As we watch, a small flock of pinkish birds flutters onto a nearby platform, expelling a couple of cardinals and immediately squabbling among themselves over the remaining safflowers. "House Finches," says Frank quietly, instantly peering through his ever-present binoculars. This is a common and familiar species around here, sometimes disliked because of their rather quarrelsome nature. Nonetheless, Frank is paying close attention, his intense gaze subjecting each individual to unexpected scrutiny. After a few minutes, Frank puts down the binoculars and promptly makes some notes in the battered notebook he takes everywhere. "Just six today, but all clean," he murmurs as he writes. "No sign of sore eyes for quite a few months now."

The reason for my visit to Frank here in small-town central Maryland is because of these House Finches, and Frank's role in the discovery and tracking of what is now one of the most carefully studied examples of a major wildlife disease outbreak anywhere. This very important phenomenon was first noticed right here in January of 1994 by Frank and other attentive bird feeders in the area. But how these "ordinary" folks became such significant players in an extraordinary story of citizens and science begins at least a decade earlier, with the advent of the Cornell Lab of Ornithology's Project FeederWatch program in the 1980s (as discussed in Chapter 4). This program was ideal for diligent and observant bird watchers like Frank, who had been systematically recording the birds visiting their feeders for years already. Project FeederWatch was a perfect way for such people to contribute their existing skills to a large-scale science project. "I was completely untrained but I did know my birds," Frank explains. "When I learned that the Cornell people were looking for regular folks to help them learn more about our birds, I got right on board. I'm quite proud to say that I was one of the original FeederWatchers."[1]

Frank signed up in 1988 and has been sending in his records without fail ever since, but he began making detailed notes long before. Nothing feathered escapes his notice, and he has often been the first to see rare and unusual species passing through. "I recorded an Eastern Phoebe and a pair of Golden-crowned Kinglets in '65, and even a Swamp Sparrow in '75. And I can tell you for certain that there weren't any House Finches in these parts earlier than 1977, March to be exact," Franks states confidently. "I was trying to attract Goldfinches with the new nyger seed at the time and was amazed to see a flock of unfamiliar finches at the feeder. At first I thought they must be slightly different Purple Finches, but the color was just not right. I had never seen them here before, but from then on we had swarms of them. And spunky! The goldies and the purples had a real hard time getting to the feeders once the House Finches moved in."

Frank had experienced the early stages of a most unusual and dramatic invasion of the eastern United States by an otherwise "native" species. Historically, House Finches were mainly dryland birds of the western region of North America. It is believed that the massive population that eventually enveloped the eastern side of the continent originated from a handful of cage birds released on Long Island in 1940. For about two decades a tiny flock managed to just hang on before mysteriously exploding in abundance around the early 1960s. House Finches spread in all directions, with the introduced birds reaching the natural range of the species in the west, evidently completing a full cross-continental distribution. Although apparently identical, the eastern-introduced finches are much more likely to be associated with humans, preferring urban areas and spending a lot of their time in gardens and back yards. These eastern House Finches love feeders, where they often aggregate in large flocks. Unsurprisingly, they are among the most common species recorded by Project FeederWatch participants; about 75% of all reports include House Finches.[2]

It was during one of his routine counts of the birds on his main feeder in January 1994 that Frank first noticed something seriously amiss with several of House Finches. Two within a small group were uncharacteristically quiet, sitting to one side, listless and immobile. When he looked more closely through his binoculars, Frank could see immediately that both birds had grossly swollen eyes that were oozing some sort of nasty secretion. "They looked to be blind and very sick," says Frank. "In fact, it was obvious that they could hardly find the seed on the platform." When he approached, the birds fluttered weakly away. He quickly rang a neighbor, Rich, also a keen FeederWatcher, and was alarmed to learn that Rich had also been seeing House Finches with crusty, weeping eyes. A few days later, Rich arrived at Frank's place with a small plastic bag and a worried expression. "It was two dead finches that Rich had found on the ground near his feeder. After some discussion, we decided to take it to our local vet. He didn't really know what was wrong with the birds but said he would send it to the university. I also let the Cornell Lab people know straightaway. But the way the birds looked, it reminded me of the sore eyes my children used to get sometimes. Conjunctivitis, the nurse called it."

These actions by Frank and Rich—and about a dozen other people from various places in the mid-Atlantic region—were instrumental in alerting wildlife agencies to the start of a significant wildlife disease outbreak.[3] Perhaps even more remarkable was that decisive action was under way very promptly. Microorganisms extracted from the tissues of the infected birds were found to be *Mycoplasma gallisepticum*, a pathogen known to affect poultry but which had never previously been associated with wild birds. And Frank's hunch was right: the disease was classified as mycoplasmal conjunctivitis, though more commonly known as the House Finch disease.[4]

Because of the efforts of Frank Wilston and other FeederWatchers, key people at the Cornell Lab of Ornithology quickly became aware of the outbreak. They soon realized that this was a highly unusual and extremely worrying situation: the pathogen appeared to have moved from domestic birds to wild species, and sick birds were using feeders and could be spreading the infection, potentially to any other species mixing with House Finches. Alarm bells were ringing. André Dhondt, who had arrived at the Lab only a few years earlier, immediately saw an opportunity for monitoring this disease using the network of existing Project FeederWatch participants.[5] Within a few months of the initial outbreak, the Lab launched the "House Finch Disease Survey," inviting Feeder-Watchers to keep a sharp eye out for birds showing signs of the disease. Usefully, these signs were all-too obvious and could be detected from a distance, and with House Finches being abundant, reporting was reliable and rapid. So was the spread of the disease. Within a year it had reached Ontario, West Virginia, and Ohio; by the fall of 1996, Wisconsin and Iowa. The rate of movement was astonishing and extremely alarming, a perception possible only because of the unprecedented geographical and temporal detail being provided by the huge number of reports flooding in. By 2004, over 10,000 participants had submitted almost 90,000 monthly reports; ten of these people (yes, Frank was one of them) had even sent in more than a hundred.[6]

This level of detail provided the framework for a broad suite of sophisticated modeling and epidemiological studies of the House Finch disease, making it one of the most intensively researched wildlife diseases in wild birds. Although much remains unclear about aspects of the disease (such as how it infiltrated into wild House Finches), André Dhondt and his colleagues continued to sieve through the data trying to discern patterns and lessons. They were able to show that the disease had a rapid and dramatic impact on House Finch abundance as it moved through an area. Within only a couple of years, overall numbers decreased by around 60% and the average group size recorded at feeders dropped from more than twelve to only about four or five.[7] This was clearly a disease that was spread easily and effectively among House Finches, though the mechanism of transmission had yet to be confirmed.

In poultry, the pathogen seems to be transmitted through direct contact between birds. Given the highly social nature of House Finches and their habit of aggregating—and agitating—in tight groups at feeders, this seemed to be the obvious explanation: cross-contamination through direct contact between birds. Direct contact would seem to be an important condition of transmission because this type of bacteria has no cell wall, meaning that it cannot survive for long outside a suitable protected environment, usually the cells of another animal. Mycoplasma does appear to be able to remain viable, however, if it is attached to certain substances that are moist and organic (objects or substrates that potentially harbor disease are known as "fomites"). Unfortunately, some feeders provide fomites in abundance: droppings, decaying seeds, moist dust, and detritus.

Thus, the pathogen could be transferred directly between birds or indirectly through fomites at the feeders.

Perhaps unexpectedly, careful study of video recordings of interactions between House Finches at feeders showed that physical contact between birds was actually quite rare. All that jostling and squabbling seemed to be birds avoiding a jab from sharp beaks; direct bird-to-bird exchange of infectious material was less likely than simple contact with the feeders themselves.[8] However, there is much more to the busy social life of House Finches than fighting for feeder space; in the cold months especially, these birds live in flocks, providing plenty of time for transmission of the disease away from the feeder. These three components of the finch-disease story—the type of flock, level of aggression at feeders, and time spent at feeders—were carefully considered in an important study conducted in Virginia.[9] By capturing House Finches at their study site and attaching minute electronic tags to leg bands, the researchers were able to track the movements and interaction of almost 200 House Finches as they visited a series of feeders fitted with tag readers. It was by far the most detailed study of feeder visits ever undertaken.

The Virginia researchers made some fairly obvious and logical predictions: the House Finches most likely to pick up the disease are those that belong to larger flocks with more stable membership, are involved in more aggression at feeders, and spend more time at feeders. Given the many possible influences and variables that were considered, these results were unexpectedly straightforward: the risk of acquiring the disease was almost entirely associated with how long they remain at the feeder. The significance of flocking and aggression was effectively trivial. In other words, the transmission of House Finch disease is clearly indirect: the pathogen is spread by the feeding structure rather than bird to bird.[10]

The Virginia study did not, however, actually examine the feeders themselves. But others have done so.[11] Swabs from a large number of feeders used by infected House Finches showed that the pathogen did indeed remain viable and infectious for up to 12 hours, plenty of time for other birds to be exposed. Once infected, the signs of conjunctivitis take a few days to develop and peak in severity at about 10 days. As the disease progresses, infected birds become inactive, often remaining on the feeder for extended periods throughout the day. This, of course, only adds to the level of contamination at the feeder. However, experimental studies found that although large proportions of infected birds develop the disease, the number of birds that died was actually fairly low, with many birds eventually recovering. Nonetheless, affected birds could remain infectious for several months.[12]

These findings indicated that House Finches visiting contaminated feeders readily picked up the pathogen before moving on to other feeders, at least for the next few days. When the disease kicked in, their lack of vigor and difficulty in seeing made remaining near or even on a feeder a sensible survival strategy. Being rendered effectively blind made the predictable location and supply of food associated with a feeder an obvious place for a sick bird to stay. On recovery, however, revitalized but still infectious House Finches were able to move on. Indeed, unlike the much more sedentary western House Finches, the eastern birds have a pronounced north–south and seasonal movement pattern, aggregating at feeders in winter—where infections are more likely—and then migrating northward, back to their spring breeding grounds, bringing the disease with them.[13]

Like almost all wildlife diseases that bring pathogens and host species together, House Finch conjunctivitis is extremely complex and difficult to understand, predict, and control. Its progression over almost an entire continent has been, however, unusually easy to track thanks entirely to the existence of an army of ideal and dedicated observers. Observers who carefully watch their feeders. Feeders where sick birds may gather. Thereby increasing the likelihood of passing the disease on to other birds. There is no escaping this conundrum: the feeders are at the epicenter of the epidemic yet are essential to the monitoring process. It is actually highly probable that the disease would not have spread if feeders had not been present, but it is also certain that we would know almost nothing about it without them.

Such quandaries are, however, somewhat vacuous: the feeders are here to stay. They may in fact provide the most effective means of future control of the disease. With infected birds apparently staying close to feeders rather than disappearing into the landscape, potential treatments (though not yet available) could possibly be administered directly to these localized birds. More immediately, the realization that feeder structures themselves may be key vectors of transmission means than feeder owners can interrupt this process simply by careful attention to hygiene (more details to come). There are extremely important responsibilities associated with bird feeding beyond the provisioning of food.

Disease Is Easy to Miss

Coming to a feeder to pick at a tightly packed volume of seed is a most unnatural way for birds to obtain food. Apart from species that feed on large animal bodies, either as predators or scavengers, almost all food sought by birds is scattered, dispersed, and spatially separated. Even when seasonal bonanzas become available, such as a tree full of berries or a hedge laden with caterpillars, there is usually room for plenty of takers. Sure, really assertive individuals may try to control access to the best bits, but such monopolies are rather difficult to maintain for long. For birds that forage largely on seeds, grain, or fruit—the majority of species visiting our feeders—their typical foraging places are characteristically in fields, woodlands, and forests. Plenty of room to avoid the crowds or prickly neighbors.

Feeders attract birds because they offer plenty of food in abundance and typically in a highly predictable location. But there are often plenty of recipients in such a deal and competition can be stiff. Still, apart from difficult times when natural foods or other feeders are scarce, picking up a snack is usually possible, even if you have to be quick. What feeders do is bring together lots of birds, of a variety of species, to the same specific spot, repeatedly and continuously. Many of these species would normally have nothing to do with one another and most of these individuals would usually avoid contact with others of the same species. Feeders can change all that, with a range of potential consequences, including the possibility of exchanging diseases.

The House Finch disease is just one example, its spectacular spread apparently directly associated with the aggregating of large numbers of birds at feeders. At a feeder the possibility of another picking up something nasty is greatly increased. Minimizing the spread of avian diseases at feeders should most definitely be one of the primary concerns of anyone engaged in wild bird feeding. This is why maintaining the cleanliness of all feeders, including the ground beneath, must always be a top priority. But illustrating this point by describing the House Finch disease story may actually be counterproductive. A key reason for the success of the House Finch disease survey was the fact that this particular ailment was so conspicuous: the birds drew attention to themselves by behaving differently, remaining at the feeder for long periods that allowed them to be scrutinized, their physical symptoms clearly visible. These features attracted attention and generated concern. Unfortunately, such visibly distinct evidence of disease is extremely unusual. Sick birds do not necessarily display their condition and, being typically lethargic, are less able to compete and are easily driven away. They are also much more susceptible to predators and are more likely to die away from the garden. This means that even feeders that are significant disease transmission sites, steadily infecting visiting birds, may not appear to be so, at least to human eyes. Only rarely do birds linger at feeders or die nearby; as well will see, the disease would need to be exceptionally virulent for birds to succumb so quickly. The problem is that the absence of overt indications of disease does not mean that none exists. We all tend to assume that because we have never seen things of obvious concern—like a sick bird—our feeders must be safe. As is becoming increasingly clear, this can be a dangerous assumption.

The Tricho-Catastrophe

The Woodpigeon is a bird usually found in rural areas of Europe though it is increasingly seen in urban areas, even on suburban feeding tables. It is also one of the most important game birds in continental Europe, with around 10 million shot annually. A significant proportion of the populations from northern and eastern Europe overwinter in Portugal and Spain, often roosting in huge numbers. Although these birds usually find plenty of natural food in the oak woodlands of the region, many also take advantage of the supplementary grain provided for other game species, such as partridges and pheasants. During late 2000, gamekeepers in a large hunting preserve in southern Spain began to notice dead Woodpigeons lying in their fields. When they started to systematically search, they found about twenty birds each day over the following 2 years.[14] By the end of the outbreak, over 2,600 dead Woodpigeons had been collected, though almost certainly many more were missed. This was of great concern because, as well as affecting future pigeon hunts, several endangered raptors occurred in the area and were known to eat pigeons, including dead ones. Subsequent analysis confirmed the pathogen to be *Trichomonas gallinae*, a protozoan parasite responsible for the disease trichomoniasis.[15] This horrifying ailment causes necrosis of the oral cavity and digestive tract, usually leading to starvation as the birds are unable to ingest their food properly. Although the disease is also found in birds of prey, it is more commonly associated with pigeons, with the main reservoir of the infection worldwide being the ubiquitous feral pigeon. Indeed, the disease has been linked to the extinction of the Passenger Pigeon in North America following the introduction of domesticated Rock Doves by the European colonists in the 1700s.[16]

A detailed review of what was the largest outbreak of trichomoniasis in wild pigeons ever witnessed concluded that the most likely vector of transmission were the feeders supplying grain for the various game species.[17] The period of pronounced pigeon mortality—2000–2001—coincided with a failure of the local acorn supply, leading many Woodpigeons to visit the feeding stations supplied for the pheasants and partridges. Of course, plenty of feral pigeons were using these places as well. Although the ultimate source of the infection was not identified conclusively, the most likely culprits were these feral pigeons, who may have contaminated the grain as they fed. Fortunately in this case—and unlike the House Finch disease—a highly effective treatment (dimetridazole) for trichomoniasis was readily available. This was subsequently added to the grain in the same feeding stations with almost immediate effect; no further dead Woodpigeons—or any other species—were detected after just two treatments. The crisis appeared to have been averted, in Spain at least. A year later, however, in the autumn of 2002, an almost identical outbreak occurred in southern England, again among Woodpigeons, but this time also among Collared Doves.[18] "Trich" was confirmed quickly, and the treatment administered. But, if indeed there were sighs of relief, they were to be short lived.

In Great Britain, concerned and vigilant bird watchers had been reporting incidents of sick or dead birds to the Wildlife Enquires Unit of the Royal Society for the Protection of Birds for years. They were also used to taking dead birds found in their gardens to the Institute of Zoology in London, where the specimens were examined by wildlife disease experts. Around the turn of the century, these specialists were particularly busy. During the period 2000–2004, a total of 750 garden birds were examined, with the majority of deaths in various finches and sparrows being associated with salmonellosis, a common bacterial ailment.[19] Although cases of trichomoniasis were detected in a few Woodpigeons and Collared Doves, the disease was not evident in any of the many finch specimens submitted. The appropriate tests had been performed for this pathogen—and many others—but had always been negative. This was no surprise. "Trich" was, after all, a pigeon problem, as everyone knew.

And then, in April 2005, a routine assessment of just another dead Greenfinch yielded a historic and deeply unsettling finding: unequivocal evidence of trichomoniasis in a finch.[20] This was the British equivalent of the discovery of the first House Finch with conjunctivitis in the United States and resulted in a similarly rapid response from a range of authorities. Remarkably, well before there had been any hint of this outbreak, the Universities Federation for Animal Welfare had brought together representatives from BTO and RSPB, various government agencies, universities, and NGOs to form a new alliance, the Garden Bird Health Initiative.[21] The goal of the GBHI was to develop guidelines explaining how people should feed birds in their gardens while minimizing the spread of disease and other risks. More specifically, they were able to undertake investigations into "the pathogens responsible for garden bird disease." They had no way of knowing what was coming their way.

Although there were a modest number of reports of finch mortality during the summer months (June to August) of 2005, the activity during the summer of the following year was utterly unprecedented. Completely unsolicited, a distraught public began to deluge the fledgling GBHI, along with many other bird and veterinary groups, with e-mails, phone calls, and small plastic bags containing dead birds. Between April and September 2006 over 6300 Greenfinch and Chaffinch "incidents" (either observations of sick birds or actual specimens) were lodged, of which 1054 were confirmed as cases of trichomoniasis.[22] What was also alarming was that these reports were coming from almost everywhere in Great Britain (though not Ireland), although the highest rates were, at least initially, from the southwestern regions around Wales and Cornwell. This was clearly a national emergency, one, again, centered squarely on the humble back-yard feeder.

To this point in the unfolding drama, all of the thousands of reports had been entirely unsolicited, the voluntary responses of distressed private citizens at finding dead and ailing birds in their gardens. Many of these people were also participants in the British Trust for Ornithology's Garden BirdWatch program (as described in Chapter 4). Like Project FeederWatch in North America, members had been diligently sending in sightings of the birds in their gardens for years; the discovery of dead finches beneath the feeder would almost certainly be something that BTO would be interested in. This network of about 15,000 households located throughout Great Britain provided an ideal workforce capable of undertaking a detailed assessment of the spread and scale of the disease. In addition, information on the disease could be combined with data on the local abundance of key species to assess how populations might be being impacted. This capacity to provide a quantitative assessment of the influence of trichomoniasis on national bird numbers was unique, something the House Finch survey was unable to provide.

At the end of April 2005, the same month "finch trich" had been confirmed, a random sample of 1614 Garden BirdWatch participants were invited to undertake additional surveillance of their gardens. These sentinel observers were carefully selected from throughout the country, providing representative samples from all regions. As well as making routine counts of birds, these people were asked to pay particular attention to three species: Greenfinch, Chaffinch, and Dunnock (the latter acting as a "control," a common species not expected to be affected by the disease). In addition to making regular searches for obviously sick or dead birds, specified signs of the disease were to be reported. These symptoms were rather generic—"lethargy, fluffed-up plumage and dysphagia (the inability to swallow)"—because trichomoniasis simply did not produce distinctive physical signs of the sort easily confirmed in the House Finches. For this reason, specimens suitable for examination were essential for confirming the cause of death. Just how this was achieved is not entirely clear. The bland, succinct "methods" sections in the scientific papers state only that "fresh carcasses were submitted by post or were hand-delivered."[23] I can't help but wonder at the reaction of the mail delivers on some occasions.

The laboratory examinations of the thousands of specimens that arrived showed that 80% of the finches submitted had indeed died of trichomoniasis, with Greenfinch accounting for the clear majority (over 62%), and Chaffinch about a third, with the small remainder made up by Bullfinch, Goldfinch, Brambling, and Siskin. (Surprisingly, salmonella, the primary cause of death so recently, was virtually absent.) This was certainly a finch-centric epidemic, and it was having a catastrophic impact. The BTO data showed that in the most affected regions of England the breeding populations of the Greenfinch declined by 35% and that of the Chaffinch by 20%, within a single year. For the whole of Britain it was estimated that approximately 500,000 Greenfinch alone had died during 2006, most during the colder months when they were more likely to be visiting feeders. This is especially noteworthy because, during the preceding twenty years (1985–2005), the national population size of Greenfinch had grown dramatically and steadily, due largely to the expansion of wild bird feeding and especially the use of nyger seed, a confirmed favorite of this species. The arrival of the trichomoniasis epidemic had, in a single year, effectively returned Greenfinch numbers to 1970s levels.

The finch trichomoniasis epidemic was unprecedented in its severity, marking the first time that a significant outbreak of the disease had been detected in species other than pigeons. The detailed investigations that followed were also the first to demonstrate that the disease could have a major impact on entire species on a national scale. The estimated mortality of 1.5 million Greenfinches for the period 2006–2009 is the largest ever associated with this disease.[24] Alarmingly but hardly surprising, trichomoniasis continues to persist in the UK and has now spread to continental Europe, almost certainly through the movement of infectious birds from Britain. It arrived in Sweden in 2008, Germany in 2009, and France in 2010. The first cases were confirmed in Austria and Slovenia in 2012.[25] The pattern of outbreaks coincides with the predicted seasonal migrations of Chaffinch. Recent genetic studies confirm that a single clonal strain of the *Trichomonas* parasite is responsible for the disease throughout this vast region.[26] Adding further complexity to this story, there had also been trichomoniasis outbreaks in North America, among House Finch (surely they have suffered enough) in Kentucky in 2002, and in the Canadian Maritime Provinces in 2007, involving Purple Finch and American Goldfinch.[27] Recall that "trich" was unknown outside pigeons and birds of prey prior to 2005. At present it is not known whether there is any connection between the disease in these two hemispheres (transmission via cage birds, for example), but in every case we are inevitably drawn to the role of the feeder.

We still do not know how or when the *Trichomonas* parasite first made its way to the previously unknown infection possibilities of finches, but most thinking has the earliest exchange happening on a typical English feeding platform during the first years of the new century.[28] The established reservoirs have always been pigeons, but the two most abundant and likely candidates, feral pigeons and Collared Doves, have been sharing the feeders with other species for decades. BTO data show that rates of visits to British gardens by these two species had been more or less stable for the preceding 15 years at least.[29] In contrast, appearances at garden feeders by the much larger Woodpigeon have increased markedly in recent years, suggesting an enhanced possibility of parasite spillover. Traditionally, the Woodpigeon has been a bird of rural pastures, where it can still be seen foraging on pasture and spilled grain in large flocks. Increasingly, however, the species has been invading towns and even large cities, taking advantage of the abundance of foods available on bird feeders. Previously, finches and Woodpigeons would have been very unlikely to interact; the bird feeder has inadvertently brought them together. The long-term consequences may be serious.

Growing Your Own Emerging Infectious Diseases

The stories associated with the detection and surveillance of House Finch disease and finch trichomoniasis are certainly dramatic and fascinating. These examples also highlight what can be achieved by well-coordinated networks of motivated citizens working with common purpose. Such highlights should not blind us to the reality that these terrible diseases are still with us, continuing to blind and starve birds, and are now well established across vast areas. Certainly the mass die-offs are less frequent, and the heat of the big epidemics seems to have cooled. Furthermore, the considerable publicity has alerted many to the importance of feeder hygiene and vigilance. Yet while we now know an awful lot about pathogen-host dynamics, the routes of transmission, and the intricacies of the particular genetic strains that are involved[30]—all essential for developing strategies for prevention and possible treatment—the really fundamental questions remain largely unanswered. How did the pathogens responsible for these diseases infiltrate their new host species? Can the diseases be treated in wild birds? There are plenty of ideas and hypotheses, but very little is certain. Our understanding of the House Finch disease, for instance, is extraordinarily detailed, yet there is still no sign of a cure. While trichomoniasis has been treated successfully in pigeons, the same approach does not seem to work in finches.

Meanwhile, the new diseases keep on coming. The latest is avian pox, a virally induced disease that causes circular wart-like growths and lesions to form on face, legs, and feet. The disease has been reported in hundreds of species globally, and it appears to persist at low levels in many locations. Most infected birds that develop mild lesions eventually recover. More severe cases may lead to impaired vision, making feeding difficult, while major lesions can result in secondary infections. Although there have been sporadic reports of avian pox from a wide variety of species throughout Europe, these have all been isolated cases and were not indicative of any sort of outbreak. Notably, although a few tit species were included, none of these were from Britain.

And then suddenly, in a pattern all too familiar by now, a Great Tit visiting a garden in Sussex, England, was found dead near a feeder in September 2006.[31] This specimen, which became the "index case" for Britain, had large, unsightly swellings covering the eyes, as well as numerous growths on its legs. Postmortem examination confirmed the pathogen to be *Aviparvirus*, the virus responsible for avian pox, and the first from a tit in Britain. This diagnosis was interesting but fairly routine at the time. However, it was soon apparent that yet another major and unexpected outbreak was under way. Again, unsolicited reports of tits with "tumors," "growths," "swellings," and "lumps," as well as many specimens, were delivered by the public to the various authorities (who had only recently been activated by the finch epidemic). The number of incidents rose rapidly from just two in 2006 to over 100 in 2010, with about 90% being tits.[32] Great Tits were by far the most common species reported, with Blue Tits a distant second, while Coal, Marsh, and Willows were all represented at much lower levels, probably in proportion to their general abundance. Of the hundreds of cases eventually reported, virtually all were from gardens with feeders. Particularly worrying was that numerous incidents involved a number of birds together, including (with photographic evidence) one alarming case where a householder had photographed diseased Great Tits and Coal Tits at his feeder, as well as a clearly pox-affected nuthatch, a species never before recorded with the disease. The first detailed assessment of this outbreak found that while the virus strain was identical to the virus that was widespread in Europe, its appearance in Britain—possibly borne by migrating continental Dunnocks, since tits do not move as much—resulted in a severe form of avian pox among Great Tits especially.[33] And the source?

> Anthropogenic provisioning of wild birds in gardens is a common pastime that influences contact rates among conspecifics and alters species complements: both factors influence pathogen transmission and exposure rates.[34]

It would be all too easy to continue describing the apparently increasing numbers of emerging infectious diseases associated with our feeders. Those mentioned already get special attention because they are new, novel, or unexpected, but there are plenty of other more routine diseases that also could be mentioned. The mainly parrot scourge known as "beak and feather disease," for example, is an exceptionally infectious viral disease of great concern wherever parrots congregate to feed.[35] It occurs naturally in many wild species in Australia and Africa but is strongly suspected of being transmitted at feeders. In Australia, Rainbow Lorikeets are a great favorite at feeders, often coming in large, boisterous flocks. Several severe though localized outbreaks of beak and feather disease have been linked, but not confirmed, with feeding (and have led to strident calls for all bird feeding to be banned outright). Thankfully, these incidents appear to have been short-lived, though the risk remains.[36]

Far more common, and therefore potentially more significant, are the bacteria *E. coli* (or to be technically correct, the strain *Escherichia albertii*) and salmonella, which causes salmonellosis.[37] Both are ubiquitous and probably global in presence but can manifest as serious diseases under certain conditions. Both have been associated with disease in a wide range of seed-eating species and appear to be highly contagious, making feeders effective places for transmission. Unfortunately, the signs of illness caused by both organisms are rather vague and nonspecific: general lethargy, a fluffed-up appearance, and a tendency to mope around feeding stations. The *E. coli* strain has been responsible for significant mortality in sparrows in Europe, caused at least one localized epidemic among North American Common Redpolls, and frequently hits British Siskins.[38] Salmonellosis outbreaks are frequent though tend to be restricted in range, affecting Greenfinch and House Sparrows most often but readily transferred to other finch species. The salmonella pathogen is particularly hardy and can persist in the environment for months. Both pathogens, unlike the others mentioned already, also pose a significant risk to humans. This means that considerable care is needed when undertaking regular—and *essential*—feeder cleaning.

When the Food Is the Problem

Of course, providing food for impaired and ill birds can also be extremely beneficial. If sickness has rendered them weak and obstructed their vision, utilizing the reliable provisions found on a familiar food table can boost condition, improve struggling immune defenses, and allow time to recover. These factors were found to have been of great value in the recovery of House Finch populations following the epidemic.[39] Without the availability of feeders, many of these birds would not have made it. Provisioning quality food on scrupulously clean feeders can be enormously important, providing sustenance while breaking the infection cycle.

Conversely, providing poor quality, nutrient deficient, or inappropriate food is unlikely to be helpful. Sick or recovering birds are much less likely to respond and could be pushed closer to the edge. In any otherwise healthy animal, a prolonged diet with insufficient energy and protein is known to seriously weaken cell function while nutrient deficiencies (especially zinc, iron, folic acids, and the B vitamins) often lead to inadequate immune defense. In other words, an inadequate diet means animals are less able to combat inevitable infections and are much more likely to succumb to disease and other problems. For example, iguanas subsisting largely on a carbohydrate-dominated diet (mainly bread supplied by tourists) had far higher internal parasite loads than those that the dominant individuals kept away.[40] In some cases, access to feeding stations actually improves tolerance of infection, allowing heavily infected individuals to survive better while transmitting even more pathogens. Indeed, there is good evidence that many birds that utilize long-term feeding stations substantially reduce their daily movements, spending much more of their time close to the feeder.[41] When significant numbers of other birds also do so, this increased density is likely to lead to an accumulation of pathogens in the local area, in turn raising the risk of transmission, even away from the feeder itself.

Poisonous Peanuts

The groundnuts universally known as peanuts were among the first "seeds" to be used extensively for the feeding of wild birds. In the pre-birdseed mix days, before the advent of ready-made packages, the easiest source of seed was the farming industry. With large amounts of corn, wheat, oats, and barley as well as peanuts readily available from produce stores and pet shops, these were convenient places to obtain food that could be put out for wild birds. Peanuts were early, consistent favorites, and remain so today, especially in Britain. These nuts—whole, in pieces, or as "hearts"—are eagerly sought by many species, and lots of people concocted their own peanut and fat balls and spreads, much to the delight of hard-pressed birds during winter. Unfortunately, many of these peanuts sold as bird food were substandard leftovers or suspect product deemed unacceptable for human consumption: the stuff that had been spilled at the warehouse, was discolored, or smelled strange. With the dramatic increase in demand for bird-feeding products in the 1980s, peanuts began to be shipped from a variety of countries around the world: Brazil, China, India, Argentina, and others. As the industry scrambled to provide sufficient supply, concern for the quality of the products was often neglected or ignored. This remains a significant issue today.

Chris Whittles, the pioneering British bird-food entrepreneur, has long been at the forefront of the "quality" wars. In the early days, he had experimented with peanuts from a variety of sources and found those from China were always favored over those from other places. "It was the higher fat content," he explained.[42] "We tested everything, and the Chinese nuts contained around 52% oils; the others were all less than 40. In winter, the birds are desperate for oils and normally try to get them from beech mast or other seeds. At the feeders, they would go for the higher oil content items every time." But the differences in the peanuts extended well beyond the oils. "Some of the stuff we saw at the docks was absolute rubbish, " said Chris. "Yet, off it went. Straight to some dodgy operator and soon to a department store near you! Worse, sometimes it was actually poisonous! I remember a large shipment of peanuts being so 'off,' it was rejected by the English buyers but was then sent to Germany. I learned later that several of the workers there who helped unload the nuts actually died from breathing the toxic dust."

Peanuts that can kill a human are almost certainly heavily infected with the serious toxic substance known as "aflatoxin." The name comes from its main source, the fungus *Aspergillus flavus*, which is present almost everywhere as spores—in the air, water, and soil—but becomes a potential problem when the spores settle and begin to reproduce. Substrates such as grains and seeds appear to be ideal substrates on which these organisms can grow, with sometimes lethal consequences for any animals consuming large amounts of infected materials. The issue of aflatoxins exploded into prominence in the 1960s when over 100,000 commercial turkeys from poultry farms in England died over a few months.[43] The mysterious illness was initially labeled "turkey X disease," though the detailed investigation that followed the emergency identified the fungus and the source of the toxins: infected peanut meal from Brazil. Subsequent testing found aflatoxins to be present in a wide range of products such as livestock and pet foods made from corn, cottonseed, and cereals, although mainly at very low levels. Research showed that most animals are apparently unaffected by mild doses of the toxins, but higher levels or even repeated low-level exposure can lead to a range of pathological and carcinogenic effects.[44] Birds seem especially susceptible, though humans are also at risk.

Because of the obvious potential for catastrophic damage to numerous industries—as well as human health (including a degree of panic about aflatoxins in unrefrigerated peanut butter)—an enormous amount of research into these organisms has now been conducted, making this now one of the best understood of all animal toxins. Many countries now have strict aflatoxin limits on grains, nuts, and seeds and impose various levels of testing. For example, both the US Food and Drug Administration and equivalent British agencies require less than 20 micrograms of aflatoxin per kilogram of all pet foods. Alarmingly, a 2006 international survey of a large number of pet foods found the worst products for aflatoxin contamination were wild bird mixes, which consistently had the highest levels, and over a quarter of these registered the toxin at 100 micrograms or higher.[45] Similarly, recent surveys of commercial birdseed mixes in Texas and the UK found between 15 and 20% of the products tested to be seriously contaminated.[46]

One of the problems with detecting this particular toxin is that even heavily affected peanuts typically do not look or even smell tainted. Without proper laboratory testing it is virtually impossible to know whether the peanuts being sold as bird food are actually safe. It is for this very reason that most of the main suppliers now clearly advertise that their products have been rigorously tested. In Britain, for example, any products bearing the logo of the Birdcare Standards Association (BSA) have been subject to careful and independent assessments using randomized scientific sampling.[47] Membership in the BSA requires mandatory testing and compliance to a range of quality control criteria aimed at ensuring the highest health and nutrition standards. In terms of aflatoxins, BSA's acceptable level is stated as "nil detectable." Despite these important initiatives, a considerable proportion—some claim most—of the peanuts and other seeds sold worldwide as wild bird food has never been tested for anything.

While a lot research into aflatoxins followed the initial mass deaths of poultry due to aflatoxin contamination, most of it focused on the potential economic and health impacts on the animals and humans involved in these industries. Tests were devised and disinfectants developed for use on farms, but there was little attention to these toxins beyond the farm or fridge. This complacency was altered abruptly with the discovery that aflatoxins were responsible for several massive mortalities of Snow Geese and Mallards in Texas and Louisiana in the 1980s.[48] The birds had been feeding on peanuts left in the fields following harvesting. When autopsies were performed on the birds, their internal organs were found to contain aflatoxins at a level the equivalent of between 250 and 500 micrograms per kilogram. In the late 1990s, corn collected from fields, again in Louisiana, where a variety of geese species had been found dead, contained levels above 8000 micrograms per kilo. In Britain, the presence of aflatoxins in wild birds was confirmed for the first time in 2006, in dead House Sparrows and Greenfinches examined at the Institute of Zoology in London.[49] Although the cause of death was found to be either predation or salmonellosis in every case, about a quarter of the sparrows and half of the Greenfinches were also found to contain aflatoxins. Given that both species are familiar garden birds, the source of the toxins was almost certainly feeders.

The implications for wild bird feeding are obvious and alarming, especially when contaminated peanuts are so difficult to identify, so much is sold untested, and, of course, affected birds are likely to die away from the feeder. A further concern is the growing body of research that indicates that prolonged exposure to low levels of aflatoxins can lead to a reduction in immune responses, especially during times of stress.[50] This could lead birds to be susceptible to various diseases or less able to cope with the rigors of winter. The deaths of a significant number of Siskins in northern England during a tough cold period, for example, were found to have probably resulted from their reliance on only slightly contaminated peanuts at feeders.[51] In milder conditions, these birds may have been less dependent on feeder food and, therefore, less likely to ingest so much toxin. We may know an awful lot about aflatoxins, but there is much we don't know about their effects on wild birds.

Selling Poisoned Bird Food, Intentionally

The aflatoxins story has resulted in many important changes in the way in which the quality of many birdseed mixes are tested and assessed. This includes the manner in which the various components are produced, stored, and transported. Protecting the truly enormous quantities of peanuts, sunflower seeds, and all the other ingredients that are the raw products of the wild bird feeding industry poses major challenges. A plethora of potential dangers abound: moisture, mold, fungi, insect pests, and rodents could all spoil or seriously damage these items, particularly while being held in storage. And as these products are intended for consumption by animals, there are very strict limits on the types of chemicals that can be used in these circumstances. Obviously, you can't just add any old pesticides to the seeds. In the United States, only pesticides approved by the Environmental Protection Agency as safe for bird foods may be used to protect stored products. There is even powerful legislation to ensure that this is the case, the Federal Insecticide, Fungicide, and Rodenticide Act, which mandates that noncompliant chemicals must be clearly labeled as such. For instance, every container of the highly effective antiweevil pesticide Storcide II, widely used to protect stored grains, has the prominent and explicit warning printed on the container: "Storcide II is extremely toxic to fish and toxic to birds and other wildlife." Nothing particularly unclear about that, you would think.

Which is why the infamous Scotts Miracle-Gro story, first reported in 2012, remains so shocking.[52] The Scotts Miracle-Gro company, based in Ohio, is the world's largest producer of lawn and garden care products, distributing a vast range of plant fertilizers and weed killers throughout the world. They are also the exclusive agent for Monsanto's deadly weedicide, Roundup. Scotts further consolidated their place in the back yards of the world by expanding into the wild bird food scene. In North America, their mixes, marketed under the names of Morning Song and Country Pride, have become popular and are widely available.

In March 2012, during an Environmental Protection Agency investigation of Scotts over a seemingly unrelated matter (the falsification of documents relating to the registration of components in lawn care products), several employees came forward to claim that the company had been adding two noncompliant pesticides—Storcide II and Actellic 5E—to packages of their bird food for at least two years. Court documents indicated that in the summer of 2007, two concerned and knowledgeable employees, a pesticide chemist and an ornithologist, approached Scotts management to warn them of the potential dangers of the practice. Despite this well-founded information, however, the company continued with the practice until early 2008. During this time at least 73 million packages of the toxic bird food were sold. The company admitted in court that they had been aware of the risks to birds. The products were finally withdrawn in March 2008 but only 3% of the packages were successfully recalled; enormous amounts of extremely toxic bird food were, therefore, distributed by unsuspecting feeders—and consumed by similarly unsuspecting birds—throughout North America. Because these chemicals affect the nervous systems of the birds, causing overstimulation and eventual respiratory paralysis, they are likely to die well away from the tainted food source. This made it unlikely that the people providing the seed would actually witness the effects of the poisons, and therefore realize that their provisioning was impacting their free-living visitors.

Not so for a Californian couple who fed their captive birds the popular Morning Song wild birdseed, purchased, as they did so regularly, from their local Wal-Mart. Of around one hundred aviary birds kept by these folks, only eight survived, most dying after a single meal.[53] Despite the official recall by Scotts and considerable media coverage throughout 2008, the seed mix responsible was obtained in January 2010, over a year after the news broke. A lot of unpalatable questions arise.

In their explanation, Scotts admitted that their primary motivation for the addition of the pesticides had simply been to protect their products prior to purchase. In other words, as one notable headline put it: "Bird food company puts shelf life before bird life!" The EPA fined the company $12.5 million in criminal fines and civil penalties, though the damage to the brand is probably incalculable. What is particularly unsettling is the suggestion, made by several commentators, that this giant corporation stopped the sale of known poisoned bird food only when its practices became publically exposed, more or less by accident. The inescapable reality of this case is that the company continued to sell what they knew to be a seriously dangerous product.

Our Role and Responsibilities as Feeders

This has been an unpleasant, and, in many ways, disturbing chapter to research and write. Although I thought that I was suitably informed about things like hygiene and the risks of disease associated with feeders, the details that emerged as I delved more deeply into these topics made me reassess the responsibilities we have as feeders—to the birds that willingly and innocently visit our tables to partake of the food we provide. By providing sustenance for wild birds, we are participating directly and possibly powerfully in fundamental aspects of their lives. The significance of this intervention—in terms of the bird's well-being (*our* well-being is another matter altogether; see Chapter 8) and physiology—will depend on many things: the bird's health, age, experience, and current nutritional state, as well as all the climatic, seasonal, and environmental variables associated with each location. There will, of course, be times when the food we offer matters much more than others; this is something that we will all probably already be aware of. When there is an abundance of natural food available, the chicks have all fledged, and the weather is fine and pleasant, our feeders and bird tables may be virtually bird free as they forage on their own away from our gardens. Alternatively, when the trees are bare, the berries are gone, the frost is severe, the rain prolonged, or the snow deep—or when an individual is weak or unwell—it is then that our feeders may be the difference between life and death. It is during these serious periods that the quality and nutritional state of the foods we provide becomes extremely important.

At the most fundamental level, all animals require an intake of organic materials in order to fuel the regular biochemical and biophysical activities associated with keeping their bodies functioning. Simply put, "nutrition" describes the process of converting chemicals acquired from outside the body (food) into less complex forms that can be used by the body's cells to perform their various functions. In other words, food is just a collection of external-originating chemicals that are consumed to enable the internal chemical reactions within the bird's body (known as metabolism) to occur. Adequate nutrition allows an animal not only to stay alive but also to move, interact, and reproduce, as well as to store some of these chemicals in the body for later use. Inadequate nutrition, on the other hand, means that key cells cannot do their jobs properly, leaving the animal unable to perform normal activities.

Nutrition can be conveniently divided into five essential components: energy, protein, essential fatty acids, vitamins, and minerals.[54] Energy is probably the most elemental of this list, being the output of cellular metabolism that provides the power required by an animal to do almost everything: move, digest, excrete, send nerve impulses, communicate, react—in a word, live. Because of this central role, the energy content of bird foods has probably received the most attention from scientists. The highest energy content of most types of food is related closely to the amount of oils and fats present. Plant seeds are by far the most energy rich of all food types. Indeed, the seeds of modern cultivars of sunflowers are among the more oil rich and hence are some of the most high-energy food items of all. Probably every bird feeder is aware of just how many birds prefer black sunflower seeds, typically known as "black oil," for good reason.

Proteins are the building materials of major parts of all animal bodies, composing the walls of all cells in tissues, feathers, and bones. As such, animals must have a continuous supply of protein in their diet to enable normal daily growth and restoration. Proteins themselves are made up of amino acids, which are either manufactured within an animal's body or acquired through the diet. Because of the importance of dietary amino acids to the functioning of the body, they are known as *essential* amino acids. The amount of protein required by an animal will vary enormously, depending on what they are doing. Young birds growing rapidly, females producing eggs, and injured individuals repairing damaged tissues, for example, all need more protein—and often quite specific essential amino acids—than birds in less taxing conditions. Most of the seeds and nuts used in bird food mixes have relatively high protein content, but this protein component is lower than many naturally occurring foods. For many birds preparing for the physiological rigors of reproduction, the best and most efficiently assimilated source of protein is found in insects. Virtually all bird species, no matter how apparently specialized their eating preference may seem to be, often include some insects in their diet. This is most pronounced, of course, in the diet provided by parent birds to their nestlings, when their tiny bodies are growing extremely fast with bones being strengthened and plumage expanded. These baby birds may grow up to be strict granivores (grain eaters), herbivores (plant eaters), frugivores (fruit eaters), folivores, (leaf eaters), or even fungivores (fungus eaters), but all will have been raised on a menu of insects as the main readily available source of protein. (Perhaps the most extreme example of a protein-powered nestling food is that of the tropical catbirds of northern Australia and Papua New Guinea whose otherwise strictly vegetarian parents feed their chicks exclusively on the fresh brains of small forest birds!)[55]

Finally, like certain vitamins, minerals of various kinds are essential for many metabolic functions. Some, such as calcium, phosphorus, and sodium, are required in reasonable amounts and are known as macroelements. Probably the most significant of the macroelements are calcium and phosphorus, which together are the primary constituents of bones and eggshells but are also involved in almost every aspect of animal metabolism. Deficiencies in calcium content, in particular, can be a major issue for captive animals but may also affect wild species as well. Insufficient calcium can lead to retarded skeleton growth, osteoporosis, and eggshell thinning, while too little phosphorus is associated with loss of appetite, rickets, and general weakness. Both of these macroelements are present in a wide range of grains, seeds, and forages such as fresh grasses and alfalfa. The optimal ratio of these two minerals in the diet for proper absorption should be calcium:phosphorus (Ca:P) around 1:1. Excessively high rations of phosphorus, however, can actually reduce the absorption of calcium, even in diets apparently containing plenty of calcium. Although most sunflower seeds have a Ca:P ratio of about 1:7, this is rarely of concern for birds consuming a varied diet.[56] Free-ranging birds are more likely to forage on a wide range of items, even though they may utilize garden feeders. Although a lot more directed research is clearly needed on this issue, the message would seem to be fairly obvious: a varied diet is best.

The British Birdcare Standards Association[57] has devised a detailed set of "Compulsory Standards for Seed Mixes" to which bird-approved food products must conform. The aim of these standards is to ensure that products sold as bird food have appropriate balances and levels of the main nutrients discussed above. The full BSA list is far too long to describe here, but minimum and maximum proportions of a large number of components are proscribed for seed mixes. For example, Group 1 items, which includes black sunflower, sunflower hearts, and nyger, must comprise a minimum of 20% of the total mix when combined. Group 2 items, on the other hand, must not exceed 80% combined; these include canary seed, millet, sorghum, and safflower. Group 3 items (19 are listed including white sunflower, linseed, buckwheat, and whole oats) together should not comprise more than 20%, and so on. Only six items are considered to be of sufficiently high nutritional value for wild birds that they can be sold as "straights": suitable to be consumed safely alone. These are black sunflowers, sunflower hearts, peanut granules, nyger, live foods, and dried mealworms.

The BSA standards also list a series of items that are "Nil Allowable": "ingredients that are not permitted in mixes displaying the BSA logo."[58] These banned items are large striped sunflowers, whole peanuts, biscuit, extruded dried pellets, all seasoned/spiced/salted ingredients, lentils, whole pulses, vetch, whole maize, flaked barley, dried rice, split peas, and barley. Other authorities have produced similar "not-to-be-used" lists that include avocado, chocolate, fruit pits, persimmons, onions, and mushrooms. I am sure there are plenty of others too. The issue here is that many of these food items are actually things people eat and are therefore readily available in the cupboard or fridge. These things are familiar and convenient—and easy to add to the bird's menu. The birds at our tables are, after all, welcome guests. Why not share our *all* our food?

Sharing food might sometimes be a significant problem. Provocatively, New York writer Jim Sterba regards the recent transition (through clever marketing) of our companion animals from simple pets to pampered family members as similar to the evocation of wild birds as "outdoor pets" by the birdseed companies.[59] The next step is obvious. Sterba's point is that if feeding chickadees is fine, why not raccoons or groundhogs—or bears? His argument is heading away from my current theme (but can be followed in Sterba's eye-opening book, *Nature Wars*,[60] though don't expect subtlety) but does return us to the important issue of our responsibilities as provisioners to wild visitors. Birds certainly consume all manner of items they find, at feeding stations or elsewhere, but that does not mean that they should. The list of forbidden items provided above is based, in part at least (some of the items are just too big and could result in choking), on nutritional standards or components genuinely harmful to birds. In many cases, most birds are unlikely to consume enough of such items to be of serious concern. Some things, however, are a real worry. Anything with salt, for example, should definitely be avoided.

Possibly the most commonplace human food supplied to wild birds globally is bread. It seems that every pond in every park in the known universe experiences the phenomenon of duck feeding, typically involving many, many slices of bread, almost daily. This is a traditional, cross-cultural, immensely popular, and worldwide interaction of humans and wildlife worthy of an entire book in itself. While the feeding of birds in gardens has now started to be the subject of serious research, we still know almost nothing about almost every aspect of duck feeding. Most crucially, despite plenty of concerns, claims, and rumors about the impact of all that bread, remarkably little can be stated reliably. Probably the most attention has been paid to ecological implications of the massive inputs of bread on the functioning of closed lake ecosystems. For example, several studies have described eutrophication and alarming levels of phosphorus in urban ponds with lots of duck feeding.[61] About the ducks themselves, however, virtually nothing.[62]

Closer to home, bread is a very common component of many garden menus. But while some promote its use, many authorities warn against it.[63] Most agree that bread provides little nutritional value and may actually exacerbate malnutrition by filling the birds with low-quality ballast. It would be nice to be definitive, but the reality is that we just don't know what effect a bread-heavy diet might have on the health and well-being of birds.

Is Meat a Meal?

Although somewhat unusual globally, the provisioning of meat for garden birds is quite typical in Australia, where large omnivorous species such as Australian Magpies, butcherbirds, and Laughing Kookaburra are favorite visitors to feeding stations.[64] All sorts of human foods are offered to these species—cheese, bread, rice, pasta, leftovers—but it is the various types of meat that are preferred (as described in Chapter 5). Thankfully, some aspects of this practice have been studied. Investigations of the blood chemistry of experimentally fed magpies found that birds consuming substantial amounts of minced meat (ground beef) or "dog sausage" (moist pet food), two commonly used feeder items, found significant increases in plasma cholesterol and fatty acids.[65] Although the health and condition of these birds appeared to be identical to unfed magpies, there are well-established physiological impacts of a prolonged fat-rich diet, including impaired liver function.

More alarmingly, however, is the possibility that birds consuming large amounts of what is usually straight beef meat may become susceptible to the calcium/phosphorus imbalance mentioned above. Beef contains about 0.01% calcium compared to 0.19% phosphorus, with a ratio of around 1:17. Were such food to dominate the diet, there is a significant chance of the birds suffering from a pathological condition known as nutritional secondary hyperparathyroidism.[66] This is associated with an inability to absorb calcium and often results in thinning of the bones (as the body seeks a supply of the mineral it can no longer obtain through ingested food) and a range of skeletal and physiological malfunctions. This particular condition is best known from captive parrots with diet limited to sunflower seeds, peanuts, and oats (calcium:phosphorus ratio of 1:10) but is also familiar to zookeepers responsible for carnivorous species such as birds of prey. Rather than feeding these birds raw meat, most zoos now include whole animals, or add skin, bone, and fur, to help balance the crucial Ca:P ratio. It is also a key reason that the guidelines for the private feeding of reintroduced Red Kites in English gardens (as described in Chapter 4) suggests that whole animals be offered rather than just meat.[67]

Concerns about the possible implications of the vast number of Australian Magpies partaking daily of the protein-rich, calcium-poor beef mince offered throughout the country were reduced when our studies showed that feeder foods made up only a small proportion of their diets.[68] Even when numerous feeding stations piled high with the stuff were just a few blocks away, most suburban magpies continued to dig up worms and grubs as they always had. Making this discovery—some time ago now—was genuinely welcome news. Confirming that at least one species that we had expected was likely to forsake natural foraging but did not was greatly reassuring. There was also some relief in finding that these birds retained their diverse diet and were not simply focused on the easy-to-find junk food option. At the time I was not aware of the potential risks associated with meat-heavy diets, so these more recent revelations about the possible implications of having a grossly unbalanced calcium:phosphorus intake was something to be reconsidered carefully. Returning to our data showed that many recently independent magpies, experiencing the realities of life away from the home territory, often congregated in numbers at some of the more generous feeding stations. These were often located in constricted gardens, typically hemmed in by hostile territorial male magpies, but represented safe havens and a source of plentiful food. Most suburbs seemed to have such sanctuaries, filled with high numbers of nervous teenage magpies, and always centered on a well-used feeding platform. Understandably, the people involved, confronted with a persistent but obvious appreciative cliental, typically respond with ever-greater piles of mince. These are the very circumstances in which the problems associated with a meat-rich—maybe even a meat-only—diet could very well be a serious problem. Whether this is the case has yet to be confirmed. What is somewhat reassuring is that these dense concentrations of magpies do not last forever. Most individuals seem to find a mate and eventually leave in search of their own patch elsewhere. And resort to the natural diverse diet of sensible grown-up magpies. I hope.

And just when you think you have seen it all, or at least been sent the YouTube clip, something entirely unexpected turns up. A journalist friend of mine, Matt Watson, recently e-mailed me a single photograph accompanied by a disconcertingly simple question: "What do you think about this?" The image had an unremarkable setting (for an Australian backyard): a typical magpie feeding station with a sizeable mound of beef. There was one glaring anomaly: the birds perched atop the pile of meat and clearly partaking with enthusiasm, gobs of mince smearing their faces, were Rainbow Lorikeets, the abundant and familiar parrots that visit my own feeder every day. To see this species, one I thought I knew reasonably well, clearly eating meat was really unsettling.

Although it started out as one of those animal curiosity pieces at the back of an online edition, the story of the meat-eating parrots rapidly acquired a virtual life of its own. Once it hit the big international outlets, especially the ultracool science website IFLS (*Vegetarian Birds Turns Carnivorous!*), it reached a vast audience, spawning plenty of great headlines of course (including *Piranhakeets*).[69] When I collated the locations and considered the over 500 completely unsolicited e-mails I subsequently received, it was obvious that this phenomenon was both common and widespread.

This provided an opportunity to investigate a completely new bird-feeding phenomenon, and I soon contacted the e-mail senders to invite them to provide more details. While this research obviously has a long way to go, a number of significant aspects have already begun to emerge. First, and very surprisingly, it has become clear that meat eating is not at all unusual among lorikeets, or, indeed, a long list of other parrots. A large number of people described pet parrots of many species being strongly attracted to meat, with tales of free-roaming birds joining the family to chew on roast dinners or steal chicken bones from plates. While a modest number of species were reported feeding on meat at feeding stations, many more were mentioned feeding on meat in the wild, usually road kill or dead farm animals. While this certainly seems bizarre, we soon discovered that almost all parrots, including those apparently specializing in nectar, pollen, and fruit, also actively seek animal food in the form of insect pupae and grubs, especially those infesting flowers and fruit.

The picture that is slowly emerging is one of birds that have always sought animal protein to add to their diet, and have done so opportunistically whenever the chance arose: insects encountered during normal foraging, a dead animal, or the discovery of an infestation of wood beetles, for example. Generally, such bonanzas will be fairly rare, unpredictable, and short-lived. Adding protein to a plant-dominated diet is essential for regular nutrition. Having access to such a typically unpredictable resource—such as a feeding station laden with fresh raw meat every day—was simply never part of the picture. But it sure is now. And at least some of these birds are making the most of it, feeding in a manner that suggests that their meat intake is dominating their daily diet. If so, this could be genuinely serious.

The material gathered in the development of this chapter has often been unexpected, alarming, and, frankly, depressing. But it is an essential and unavoidable part of the story. The act of feeding birds can have serious consequences. Some of the risks are now clear. But feeding can also be profoundly important, actually playing a role in preventing extinction. To consider this aspect of bird feeding, let's head far, far south.

7

Feeding for a Purpose

Supplementary Feeding as Conservation

It's a long way to New Zealand from just about everywhere, though thankfully, not from Australia. For such a remote and relatively small collection of islands almost at the bottom of the planet, its place in the minds of birders and conservationists is distinctive and rightly celebrated. Tales of the decidedly odd kiwi, sheep-eating parrots (Kea), and those gigantic though extinct moa are more than enough to fire the imagination of any keen naturalist.[1] All too often, however, the reality for visitors seeking to encounter the remarkable endemic birdlife of Aotearoa (the increasingly used Maori name for the country) is typically disappointing. Since the arrival of humans, first the Polynesians and, more recently, the European settlers have been extraordinarily successful in transforming a land of dense, dark forests into an almost exact replica of genteel rural England. And that includes thriving populations of European birds, many declining or rare in the UK and Europe. Wander through the delightful parklands in Auckland, Wellington, or Christchurch and Chaffinches, Song Thrushes, and Greenfinches are everywhere. Set up a feeding station—as demonstrated so clearly by Josie Galbraith—and it will soon be swarming with House Sparrows, Blackbirds, and Starlings.[2]

But where are the native species? Certainly, some are there in the gardens and parks as well—small ones such as Silvereye (or White-eye), Tui (a vibrant, vocal honeyeater), and Grey Warbler—and even larger birds such as Kererū (the huge fruit- and leaf-eating New Zealand pigeon) and Kākā (a close relative of the Kea parrot of the mountains) are increasingly being seen in towns and cities, a direct result of habitat restoration and, to some extent, garden feeding. But these successes are limited though welcome exceptions to the conservation status of many New Zealand birds. Despite the extraordinary efforts of many passionate New Zealanders, their birds are in serious trouble. There is no mystery to this terrible state of affairs. Starting with the arrival of the first humans only about 800 years ago, several devastatingly effective processes were unleashed: the mass clearing of the dense temperate forests, unusually efficient hunting by people, and the impact of the mammalian predators that accompanied them. The destruction of huge swaths of forest and the rapid extermination of the nine species of moa by the Polynesian predecessors of the modern-day Maori are disturbingly well understood; these were events that occurred relatively recently, allowing researchers to piece the puzzle together all too well.[3] After all, New Zealand was the last major landmass on earth to be colonized by people, allowing the waves of calamity to be mapped and dated with alarming precision. Until around the 13th century—a time when King Edward I reigned in England and the Mongols were raiding central Europe—these mist-shrouded islands, stretching for 1200 kilometers (745 miles) across the cold southern ocean, remained unknown to humanity. Their remoteness and isolation from all other lands allowed a remarkable community of animals and plants unlike anywhere else to evolve, most significantly because an entire major component of the vertebrates was missing. When the bits of land that were to become New Zealand split off from the original supercontinent of Gondwana, there were no mammals onboard. As a result, over the following 80 million years, these islands became the land of birds, diversifying into around 200 species and occupying every possible niche from the dark floor of the dense podocarp forests to the tops of the soaring snowy mountaintops.[4] There were minute wrens seeking insects along the coastal beaches, formidable Adzebills (a huge rail) striding through forests, and, of course, all those moa, ranging from goose-sized to the towering Giant Moa that could stretch its neck to over 3 meters (10 feet) in height, taller than any other bird. Remarkably—and very significantly—a large proportion of the land birds of New Zealand became flightless.

This propensity for New Zealand's birds to stay on the ground instead of flying was not simply because there were no predators. The broad diversity of lifestyles that evolved included plenty of carnivorous species including those Adzebills and Haast's Eagle, the largest bird of prey in the world.[5] Some of the moa were also predatory, but it appears that for most birds, running, hiding, or freezing were the best means of survival. Avian predators tend to be visual, relying on movement to detect their prey. Mammals, in contrast, smell out their potential meals. You can tell where this story is heading.

When those first waka (the massive Polynesian ocean-going canoes) crunched onto the pebble beaches of the northern extremities of New Zealand,[6] it was not only the stone axes, firebrands, and hunting spears accompanying the people who disembarked (into what would have been a bewilderingly different world) that were to utterly transform this land. Also aboard, either as typical ship-dwelling vermin or possibly as fresh meat for the long voyage, were Kiore, the Polynesian rat, a species found throughout the islands of the Pacific.[7] Their first encounter between a New Zealand bird—say a North Island Snipe wandering along the nearby dunes—would have been predictably brief and fatal. The complete lack of fear the bird would have probably shown, frequently reported as innate stupidity or beguiling innocence, is better explained by the mammal-free eons of isolation. Nothing had prepared them for the fur-bearing catastrophe that was about to engulf them. As is continuing to happen today, the bird would have simply stopped where it was, trusting in its camouflaged plumage. The Kiore is only a relatively modest-sized rat (40–80 grams [1–3 ounces]), but it appears to have wiped out or greatly reduced many of the land birds within a remarkably brief time.[8] Perhaps most significantly for the overall ecology of the country, the rats appear to have ravaged massive colonies of ground-burrowing petrels, forcing entire populations to abandon traditional breeding grounds and relocate to offshore islands. According to recent analyses, the sudden cessation of millennia of guano production by unimaginable numbers of seabirds would have had enormous and long-term consequences for the fertility of vast areas downhill of the colonies.[9] It seems hardly possible, but the arrival of a small rodent hundreds of years previously would profoundly influence agricultural production in the future.

While the work of the kiore would have happened unseen in burrows and in the dark of night, the ruthless efficiency of the daytime Maori moa hunters was proceeding apace. Again, it appears that the lack of fear of humans made the moa pathetically easy to hunt and kill, at least until inevitably they became harder to find. Enormous butcheries consisting of the bones of vast numbers of moa attest to the scale of the bonanza; one site alone (Shag River on the South Island) accounted for over 6000 individuals.[10] Of course, such protein riches were not to last; some authorities have suggested that moa meat fueled considerable growth among the Polynesian populations and that the collapse of this resource had major consequences for Maori society and culture.[11] While the social implications of these events continue to be debated, the ecological impact of the rapid loss of what would have probably been the most influential players in the complex bird community can only be guessed at.

And that was only Act 1 in the unfolding tragedy of Aotearoa. The arrival of the first Europeans, possibly starting with the very first ship in 1769, was to dramatically accelerate the changes already under way. The steel axes, fire-making flints, and muskets and especially the Eurocentric perspectives of the white-skinned colonists soon transformed the appearance of the countryside. Again, however, it was the intentional and accidental mammalian introductions by the Europeans that were to have truly catastrophic outcomes for the birds of New Zealand. The House Mice, Norway Rats, and Ship (Black) Rats would have simply jumped ship, as normal, while some domesticated pigs, dogs, goats, sheep, horses, and cats probably escaped from farms and homesteads to start feral populations in the wild.[12] Thus has been the outcome of virtually every European colonial endeavor around the world.[13]

What has been especially noteworthy about the New Zealand example was the ambitious and imaginative range of species intentionally introduced for a variety of dubious reasons. While not all have survived (and ignoring over thirty species of foreign birds now established), New Zealand is now home to 34 mammals from elsewhere, from Himalayan Tahr and elk to hedgehogs and hares.[14] For our present purposes, however, it is those that have taken to eating birds that are most important: the rodents

(House Mouse, Ship Rat, Norway Rat, as well as the Kiore), cats (domestic pets now feral) the mustelids (ferret, stoat, and weasel) and, perhaps surprisingly, the Australian Brushtail Possum. The latter, introduced to establish a felt industry for hats, have proved to have a major impact on eggs and nestlings.[15] All are ferocious predators, though two, rats and stoats, appear to be particularly effective, and widespread, even swimming freezing seas to reach certain islands.

The result of all these influences has been a biodiversity in free fall. Because these events have been so recent in New Zealand, it has been possible to date the disappearance of each species lost over the past couple of centuries with horrible precision. A recent book covering the extinctions of New Zealand's birds provides an alarming graph of the stepwise disappearance of each species, shockingly arresting in its precision and finality.[16] From the original 223 species known to be present prior to the arrival of humans, an abrupt plunge downward with the arrival of the Polynesians, the notation stating baldly, "All moa, geese, Adzebills and Haast's Eagle [disappear]." Twelve species gone. The downward steps continue, gradually becoming a smoother curve declining increasingly steeply from about 1800, the labels now thickly clustered and unrelenting, starkly documenting annihilation: North Island Kākā (1851), Hutton's Rail (1893), Laughing Owl (1914), Bush Wren (1972). Unique species, gone forever, and many others on the same trajectory. Yet it seems that the bitter reality of this predicament—the very real possibility that even more of their precious animals could yet disappear—has transformed many New Zealanders into some of the most imaginative, innovative, and resourceful conservation biologists and managers on this planet. Facing extinction daily seems to have focused the mind and steeled their resolve; finding, slowing, and stopping the mammalian predators was paramount.

While the traditional approaches of trapping and poisoning have been significantly enhanced and fine-tuned (fortuitously assisted by the lack of native mammals), it soon became evident that eradicating these pests was simply an impossibility in such a large country with extremely challenging terrain and a human population of only about 4 million. One solution was to keep the predators away from the birds by physical separation. New Zealand has an abundance of islands just offshore, and the rather obvious idea of taking vulnerable birds to these places was tried as early as 1895. In a remarkably prescient and risky exercise for the day, a large number of birds were captured and released on islands in Dusky Sound at the bottom of the South Island in a desperate attempt to secure populations threatened by predators.[17] Unfortunately, this farsighted venture failed when stoats reached the islands only a few years later. This approach was, however, revisited around the 1970s as methods of detection and control of mammals became far more effective. Although it took some time to perfect, the ability to actually remove all unwanted species from an entire island meant that the New Zealand Wildlife Service (subsequently the Department of Conservation (DoC)) was able to return to serious island translocations. DoC's skills at removing every last possum, stoat, and rat and restoring the original vegetation rejuvenated the country's desperate interest—and hope—in saving their precious birds. By the 1980s, this "island sanctuary" approach was to become a mainstay for the conservation of New Zealand's birds. Today, numerous islands dotted all around the country have thriving and healthy populations of many iconic species.[18] For obvious reasons, access to most of these critically important island reserves is highly restricted, and many are remote and inhospitable. There is, however, one major exception.

Visiting a Real Ark

To get to the island of Tiritiri Matangi, you don't need to wait 3 weeks for a special DoC temporary visitors permit, arrange to be accompanied by a ranger, hire a helicopter, and come prepared for possible blizzards or ice storms. Instead, you can wander down to the harbor terminal in the Auckland CBD and catch the daily ferry. The brief voyage provides spectacular views of the city towers and the sails of hundreds of yachts against the massive glowering cone of Rangitoto, the supposedly dormant volcanic island only 8 kilometers from New Zealand's largest city. I went one startlingly bright day in December.

Directly ahead, the island is coming into view, its dark green forests contrasting with the white sandy beaches lining the shore. I am on my way to Tiritiri Matangi to see for myself some of the results of the extraordinary results New Zealanders have achieved in saving their birds, and to learn more about the role that supplementary feeding has played in these programs. The provision of additional food plays a major role in conservation and wildlife management plans throughout the world, used for a vast number of species and situations.[19] Describing the diversity of these programs is well beyond the scope of this book. Because of the depth of research associated with these activities, however, there is much that we can learn that may be of interest and value to bird feeders in more domestic circumstances. Are there lessons from well-studied examples of applied supplementary feeding for conservation and management that might inform garden feeding?

My guide is Josie Galbraith, a colleague who has been researching bird feeding in the suburbs of northern Auckland, just visible behind the ferry in the sparkling sunlight. As well as an accomplished researcher in her own right, Josie is the ideal companion to have on this trip to Tiritiri as she spent much of her life visiting the place. "Multiple times every year, since before I was born," she claims. Her father, Mel Galbraith, an ecologist from Unitec in Auckland, was directly involved from the earliest days in the complete transformation of the island from overgrazed rocky sheep paddock to vibrant restoration of the original New Zealand landscape. When the restoration work started in the early 1980s, only 6% of the 220 hectare (540 acre) island was covered in trees. Today, after the establishment of almost 300,000 trees, 60% is now forested, with the remainder intentionally maintained as open grassland.[20] Since toddlerhood, Josie has been helping in the huge amount of digging, planting, weeding, trapping, and all the other exhausting practical tasks necessary for the restoration of a complex ecosystem. "An army of very hard-working volunteers spent years, usually over their weekends, doing all the work," explains Josie. "I still think it was a remarkable effort by these people because no one really knew whether it would actually work." There are a lot of steps between the theoretical concept of restoring an entire island and the reality; even with all that effort there were no guarantees. It did come off—spectacularly! Even as the vessel slows down to dock alongside the island's only jetty, we are greeted by sounds almost lost forever: the haunting tapestry of strange and unexpected noises that demonstrate in the most exuberant way possible the reinstatement of a New Zealand bird community.

A sign—"Tiritiri Matangi Open Sanctuary"—greets us as we step off the ferry, an explicit proclamation of the island as a place where visitors are welcome (although limited to a maximum of 150 per day); most of New Zealand's other conservation islands have extremely restricted access. Before we move away from the jetty, a Department of Conservation officer welcomes us and explains a little of the history of the island, the extraordinary efforts that have gone into its restoration, and the critical responsibilities associated with keeping the continuing threats at bay. Undoubtedly, the biggest concern remains that of ensuring that mammalian predators do not reestablish. Remarkably, even though the island is only a few kilometers from the mainland, the only invasive mammal present before restoration was the Kiore, which was finally eradicated in 1994. But rodent invasion remains an ever-present threat; before the ferry left the Auckland terminal each passenger was required to check their baggage for rodents. It seemed a little extreme, but it was an appropriate indication of the seriousness with which the island's stewards take their responsibilities. The challenge of keeping Tiritiri predator free is formidable, being such a short distance from a major city full of the usual vermin. So far, so good.

I had hoped that I might catch at least a glimpse of some of the species I had previously seen only in captivity but was not expecting this to be so effortless. As the track leads away from the beach into the dense restored forest, I am astonished to see plenty of Red-crowned Kākāriki (parakeets), Kererū (the large endemic pigeon), Kōkako (a strange crow-like endemic) and Tīeke (also known as Saddleback, another odd endemic) foraging in the foliage just above me, the birds apparently oblivious to the presence of overexcited humans. Their legendary lack of fear was a fatal flaw of many New Zealand birds, but it certainly makes the birding easy. It is also immediately obvious that these are not simply wild birds in the bush; almost all wear colored leg bands, direct evidence that, as well as being a sanctuary, Tiritiri is a functioning science laboratory. "Most of the translocated species are being intensively researched," says Josie, who has been involved in many ecological studies here. "This entire exercise is an ongoing experiment, and we need to learn as much as we possibly can. For example, the first of these birds were brought to Tiritiri from other parts of the country. But would they find the resources they needed to survive and breed? What would they eat?"

Further along the track we begin to head downhill into a steep valley where the forest is particularly dense and dark, the entire area overshadowed by the huge spreading canopy of several enormous and obviously ancient trees. "These are Pōhutukawa, the oldest trees on the island," explains Josie. "Some of the very few to have survived all the clearing. They must be hundreds of years old." The path is carpeted in their rich red flowers, a rare element of color amid all the dark greens. As we pause to admire these venerable trees, I am a little surprised to detect that telltale flitting behavior of birds visiting a feeder. Sure enough, just ahead is a box-like platform with a peaked roof supported on a pole. About a dozen or so small birds are dashing in and out, landing on the roof or entering the mesh box through a series of small holes. I recognize Tui and bellbirds, both fairly common honeyeaters found throughout New Zealand, but what I am particularly excited to see are Hihi (also known as Stitchbird), another a nectar-feeding species and one of direct relevance to this trip. Although drab from a distance, up close the main fawn-mousy brown plumage of the male is offset by striking yellow, black, and white. This is another of the really odd New Zealand birds, an ancient endemic species recently placed into a taxonomic family all by itself. The Hihi has also become famous among behavioral ecologists because of its highly non-traditional practice of copulating face-to-face, something unique among birds.[21] But whatever their preferred sexual position, it has become clear that the really essential ingredient for successful reproduction in Hihi is sufficient sweet stuff: the rich carbohydrates normally supplied in the nectar and fruit these birds crave. These treats are abundant in the places where Hihi have survived but appear to be severely limited elsewhere. Including Tiritiri Matangi.

Helping Hungry Hihi

Along with so many other New Zealand birds, Hihi were almost entirely obliterated by the gray tide of introduced mammalian predators.[22] As early as the 1880s, Hihi had been exterminated from the mainland, with the entire population limited to those present on the mammal-free Little Barrier Island.[23] This thickly vegetated and mountainous island of 3000 hectares (7400 acres) located to the north of Tiritiri harbors numerous species that had either been exterminated or greatly reduced elsewhere and is therefore an extremely important sanctuary for many of New Zealand's beleaguered birds. But Little Barrier is also just one small island, and because it carries the fate of so many vulnerable species, the risks of something going wrong are always high. A major storm, a wildfire, a virulent disease—or the arrival of rats or stoats—could have unthinkable consequences. To reduce this risk of extinction as well as contribute to ensuring the survival of many species, Little Barrier has served as both a continuing safe haven as well as a source of birds for translocation. In 1995, for example, the first group of Hihi was brought the short distance (about 50 km, or 30 miles) from Little Barrier Island to Tiritiri Matangi.[24] By then Tiritiri had been thoroughly and carefully revegetated; it was quietly assumed that these birds would find their new home to their liking. After all, several other species had also been brought over to Tiritiri and were thriving.

The Hihi, it was discovered, are somewhat choosy when it comes to nest locations, preferring natural hollows, which were simply much rarer in the relatively young forests of Tiritiri. This was easily solved with the provision of lots of nest boxes; the birds took to these quickly and eggs soon appeared.[25] Less straightforward was their sugary diet. While Tiritiri did have plenty of the nectar-bearing plants that Hihi like, the abundance of flowers and perhaps the quality of their carbohydrates seemed to be of a lower quality than those back on Little Barrier.[26] What was much more obvious, however, was the level of vigorous competition over these sugar supplies. The presence of the much larger and more assertive Tui, as well as the sheer number of other "honey-eating" birds such as bellbirds, suggested that the Hihi were losing the competition for the limited supplies of sweet food.[27] And without sufficient suitable foods, breeding was almost always unsuccessful, if it occurred at all. With the survival of Hihi being increasingly reliant on translocation to islands such as Tiritiri Matangi, this was an issue of direct and desperate importance.

Provision of supplementary foods was called for, although the type of food and the timing were not clear. Various concoctions and feeding regimes had been trialed at different locations, but careful monitoring of the birds found little evidence of success. A crucial breakthrough came with the conclusion of a 6-year experimental study of Hihi translocated to another island, Mokoia, which is located within Lake Rotorua in the famous geothermal region of the North Island of New Zealand.[28] In research lead by Isabel Castro from Massey University, several different high carbohydrate supplements (supplied in modified hummingbird feeders) were positioned near the nest boxes of randomly selected breeding females, intentionally favoring the nesting bird by easy access to the goodies. "Unfed" females had to fly over 100 meters (110 yards) to visit a feeder and were unlikely to find any food left over. During the study, treatments were switched and feeders moved so that reliable comparisons could be made. In particular, the researchers were interested in assessing whether providing supplementary foods throughout the entire breeding season was important, previous experiments having been limited to only 2 weeks of feeding.

The first thing that the researchers noted was that the birds really did use the feeders a lot, and did so while incubating eggs and brooding chicks. Although they had to travel far greater distances, even the females without feeders nearby also visited feeders if they were able to. There was no doubt that the additional foods were being sought after. But the critical question was whether this made a difference in their reproduction. Having reviewed supplementary feeding experiments in Chapter 5, we would probably expect some effect, but what Isabel Castro and her colleagues found was almost entirely unexpected—though very welcome. Females with access to the supplementary foods actually produced more eggs than those that did not, and more than doubled the number of young successfully fledged.[29] For Hihi, the additional food was not simply useful; it was the difference between survival and death or extinction. This stark conclusion was reinforced by the treatments without supplementary provisions: these birds simply were unable to breed successfully. These were remarkable findings—very few feeding experiments have altered clutch size—and of immediate relevance to those concerned with Hihi conservation. Given that almost all the locations being considered for translocation are recently restored islands (including fenced reserves on the mainland) with similarly limited food supplies, providing supplements more or less continuously seems to be essential for Hihi for the foreseeable future. Hence the feeders on Tiritiri and all the other places where Hihi have been reintroduced. For this special species, survival requires continuous intervention.

The View from Tiritiri Matangi

It is about noon and fairly warm when Josie and I decide that it is time for a lunch break. We have emerged from the forest onto a grassy headland overlooking the deep blue-black ocean on the northern part of the island. A raised embankment covered in soft, thick grass offers an ideal place to pause and extract food from our day packs. "This is a great spot," declares Josie, with just a hint of mystery. Gazing out through the sea mists to the northwest I can just make out the shape of what is obviously a large, mountainous island. "Little Barrier Island!" exclaims Josie quietly, as though simply naming the place conveyed some impression of its significance. Before I can respond, however, we are confronted with an immediate and urgent bird-feeding emergency: someone has just stolen Josie's sandwiches!

Throughout this visit I had been expecting to see a Weka, the small dull-colored rail species that is, without doubt, the most adaptable and adroit of New Zealand's native birds at exploiting any possible foraging opportunity. I suspect many visitors to this country can relate stories of the cunning and resourcefulness of this little rail, typically involving the stealing of food from right under their noses. Obviously, I thought, the Weka strikes again! Except, as Josie informs me, Weka do not occur on Tiritiri. . . .

Instead, the rail that strides confidently out from the nearby dense undergrowth is not small and gray but alarmingly large, deep blue and rich green and, for me, entirely unexpected. This is a Takahē, undoubtedly one of the world's rarest birds (total world population about 200) and one of the species I most hoped to see on Tiritiri. Josie had been circumspect about my chances, however, explaining just how unpredictable sightings could be. "After all," she had said only a few minutes previously, "there are only 9 on the entire island. Although this particular spot is often pretty good . . ." Josie planned this potential encounter nicely, though she had not foreseen the loss of her lunch in the process.

This is an extraordinary experience! I am lounging on a grassy bank on a beautiful island a few meters from one of the rarest birds on earth, a species on the very precipice of extinction. Despite seemingly overwhelming odds, however, this extraordinary species remains vibrantly alive, its prospects of long-term survival steadily increasing. The Takahē is a truly massive rail, at about 3 kilograms (6.6 pounds) by far the largest rail in the world. (Although the Moho, a now extinct New Zealand species (last seen in 1894), was even larger.)[30] Seeing a Takahē so close, one cannot fail to be impressed by their robust, solid build, no-nonsense demeanor, and massive red industrial-strength beak, capable of snipping deftly through coarse vegetation (or fingers, I suspect) with ease. I had been aware of the rescue of the Takahē for some time, having read about the painstaking effort taken to understand its ecology and behavior, and knew it to be one of the species released on some of New Zealand's predator-free islands.[31] Yet all these facts simply had not prepared me for the living, breathing reality of the creature up close. And this stridently confident super-rail is just one example of the many species almost certainly still extant because of islands like Tiritiri Matangi.

This particular wildlife-human interaction concludes abruptly with the Takahē disappearing suddenly into the dark undergrowth nearby, having discarded Josie's plastic lunch box unopened. It seems that our bird, unable to procure anything edible, continued on with other pressing business elsewhere, such as ambushing unsuspecting picnickers farther along the path. If so, would these people be able to resist? "It's just a cookie? What difference would that make?" This is an all-too familiar scenario, played out in picnic spots every day, the world over. What may seem trivial and commonplace in a suburban park, however, assumes a special pertinence here on Tiritiri. Ensuring easy access by anyone to one of New Zealand's outstanding conservation islands has been a fundamental goal of the custodians of the island from its very beginning. As I had just experienced myself, encountering the living evidence of successful conservation in the form of species now effectively impossible to see in the wild can be profoundly moving. The other side of this easy accessibility, however, is the risk that visitors could become complacent about what is required to retain the conditions necessary for conserving the birds. As readers of this book are likely to appreciate, resisting the temptation to offer something to nibble to an eager animal can be very difficult. But feeding wild birds, like Takahē, in a place like Tiritiri Matangi—or any conservation reserve—can be a serious matter indeed. Sometimes even apparently minor changes to a bird's diet can have a significant influence on their lives, especially when it comes to breeding.

Understanding the detailed relationship between food and reproduction has been fundamental to some of the desperate conservation rescue stories played out in New Zealand. Perhaps the most desperate—and celebrated—of them all is the case of the Kākāpō, the famously odd parrot, an important part of which took place on the misty island our eyes are repeatedly drawn to on the northern horizon.

Little Barrier Island is entirely different from its neighbor, Tiritiri Matangi. Where the latter has a gentle terrain and a welcoming manner and is open to all after a brief ferry trip, Little Barrier is brutally rugged, demanding, and unforgiving. Access is also stringently restricted; even for those authorized to visit, extreme fitness and resolve are necessities. I have long dreamed of making such a trip but, so far it had been just too difficult. Josie has, however, been to Little Barrier several times as part of a conservation team. As we wander slowly along the cliff-top track, the misty mass of Little Barrier just evident on the horizon, I am keen to hear her experiences. She needs little encouragement.

"I have been over several times recently, trying to catch [Red-crowned] Kākāriki [parakeets] for translocation," Josie explains. "Little Barrier feels so, so ancient as it looms out of the sea as you approach on the boat, feeling very much like a scene from *Jurassic Park*. It does not look at all inviting, with its dark, thick forest cover, steep cliffs, and the cloud clinging to the summit. The beach is comprised of large boulders, which makes landing a boat very hazardous, although this has probably helped to protect the island from casual visitors. And you can see and hear the birdlife before you even land: Kākā [one of the large parrots] screeching and whistling, Tui chasing each other, Kererū [New Zealand pigeon] performing their aerial dives. Once you are safely on the island there is a real sense of what this country must have been like before people. The forest is a bustling, busy, noisy place both day and night, full of birdlife. It is always striking to see the huge flocks of 50, 60, 70 or more Kererū foraging on the ground, something that just doesn't happen on the mainland anymore." These are sights and experiences possible only because Little Barrier has none of the marauding mammals that have caused so much damage elsewhere in New Zealand. In their absence, as Josie has so poetically described, it is possible to maintain the ancient atmosphere of Aotearoa as well as providing a safe refuge for some special guests. Which is why this journey exploring bird feeding has brought us—metaphorically—to the resounding forests of Little Barrier Island, trying to feed a very strange parrot. It is a provocative illustration of the importance and risks of employing supplementary feeding for specific purposes.

Preparing a Smorgasbord for a Peculiar Parrot

There are plenty of contenders for the title World's Strangest Bird, but by any measure the Kākāpō must surely come close to the top of the list. This mossy green monster is easily the largest parrot by weight (at

3.6 kilograms (8 pounds), heavier than the biggest macaw) and is, not surprisingly, entirely flightless. It is also nocturnal, the only night-going parrot. It exudes a very strong, musky odor that has been likened variously to honey, the scent of fuchsias, and, rather imaginatively, the inside of a clarinet case.[32] When it comes to reproduction, however, things get really odd. Unlike most parrots, which form close, sometimes lifelong pair bonds with their mates, Kākāpō do romance rather differently. Every evening during the long summer breeding season, the males trudge up to the top of a prominent forested hilltop or ridge where they settle into a dusty depression that they (or their predecessors) have slowly excavated over a prolonged period, these well-worn mating grounds being known as a "track and bowl" site. Within this bowl, prone and primed, they slowly inflate their huge frontal air sacs (which expands their body into a circular pillow shape) and begin to "boom," forcing the air out through their nostrils to produce a weird, low-frequency, and extremely unbirdlike call. These vocal displays can continue, more or less continuously, throughout the night, the strange low moaning sounds traveling far off over the surrounding landscape.[33]

The bizarre display is intended to advertise the presence of the males to any prospective females within earshot, who are theoretically free to choose among the various boomers available in the local vicinity. This type of mating arrangement (a "lek," after a Swedish term for "play"), where males perform and females select, is well known among species such as bowerbirds, sage grouse, and birds of paradise, but it is practiced by only one of the 350 species of parrot.[34] Should this elaborate process be successful, the birds mate ("overly vigorously," according to one early observer) before the female departs to prepare a nest; she has no further interaction with the male. The nest site is typically located far from the male's calling bowl and placed under a dense thicket, presumably as protection from the typically wet and windy weather. She remains at the nest for over 100 days, and because she incubates without assistance from a partner, is obliged to leave the nest and eggs unattended whenever she needs to forage.[35]

Noisy, smelly, heavy, slow, flightless, remaining for prolonged periods in predictable locations, and relying on camouflage and stillness to avoid detection—it is hardly a mystery why the Kākāpō was especially vulnerable to the relentless tide of mammalian marauders, especially stoats.

Having previously been found throughout the country, and often in abundance, by the 1880s the only remaining Kākāpō were restricted to remote mountainous regions and Stewart Island, the third island of New Zealand in the extreme south.[36]

The Kākāpō story—the realization of catastrophic demise, the early and heroic attempts at rescue, the repeated heartbreaking failures, the pitiless pressure of the invasive predators, the ever-present likelihood of extinction, and the recent breakthroughs—all make for one of the world's truly great if nerve-racking conservation stories. The details are far too complex and intriguing to be given justice here but have been well described elsewhere (most eloquently in William Stolzenberg's *Rat Island: Predators in Paradise and the World's Greatest Wildlife Rescue*).[37] Suffice to say, despite astonishingly farsighted attempts in the 1880s to establish populations on predator-free islands (over 400 Kākāpō were moved to Resolution Island, only to have stoats reach the island within a few years) and plenty of other desperate actions, the population continued to slide. A long and devastating period of attempting to establish captive breeding facilities failed utterly, with no birds surviving longer than 4.5 years and not even a hint of reproduction. When you recall how these birds go about their courtship, this is hardly surprising.

By the 1970s, the accelerating process of removing the mammalian invaders from various offshore islands provided opportunities for the transfer of Kākāpō from the last surviving population in the country.[38] These were fraught times. At the time it was believed that only about 20 birds remained alive, all in the extremely remote and rugged mountains near Milford Sound in Fiordland on the South Island. Every aspect of this exercise was extraordinarily difficult and inherently risky: the finding of the birds, their capture, containment, transport, release—plenty could go seriously wrong at every step. Yet the risk of doing nothing meant almost certain extinction. The first transfers began in 1974, all the way to tiny Maud Island in the Marlborough Sounds at the top of the South Island. Over the following few years a total of five Fiordland Kākāpō were found, captured, and released, apparently a quarter of the entire population. Unfortunately, it was later discovered that these were all males. Reproduction was not that likely.

And then, came the startling discovery, in 1977, of an entirely unknown population of Kākāpō on Stewart Island, far to the south.[39] With estimates of over 100 birds, this was wonderfully welcome news and prompted the remarkably resourceful New Zealand Wildlife Service to revise their Kākāpō conservation strategy to include both safeguarding the Stewart Island birds and continuing translocation to suitable islands. The discovery of a population of this size also provided an opportunity to conduct some fundamental research into this peculiar and very poorly understood species. In particular, little was known about the Kākāpō's diet, a potentially critical influence on breeding activities. The fragmentary information available on Kākāpō reproduction indicated that breeding occurred erratically and was rarely annual. Surely some aspect of the food supply was implicated and could hold the key to understanding what led to breeding. Studies of the Stewart Island birds showed that they chomped their way through all manner of foliage, stems, roots, buds, fruits, and seeds—seemingly any vaguely edible part—of a wide variety of local plant species.[40] While these foods were somehow capable of supplying the nutrients required for daily survival and maintenance in this bulky bird, it was unlikely that most of these items would have supplied the key resources, especially the protein, required by females to breed.

A major breakthrough (well, confirmation really, as these insights had been first made by naturalists in the early 1900s) came when it became clear that the periodic breeding activities of Kākāpō on Stewart Island almost always coincided with the masting events of two of the predominant tree species on the island.[41] Known as Rimu and Pink Pine, these large podocarp trees produced huge crops of seed every 3 or 4 years, which showered the forest floor beneath the trees. The crude protein levels of these seeds provided a diet with almost twice that of the birds' regular intake.[42] Although the trees produced some seeds annually, it appeared that the birds needed the protein in bulk to get their reproductive systems moving as well as supporting chick rearing.

Meanwhile, the number of Kākāpō known to be alive continued to decline, including the vital residual population on Stewart Island, due to predation by feral cats and the birds' naturally low breeding rate. By the late 1980s, with only about 40 birds left, the monumental decision was made to relocate all of the remaining Stewart Island Kākāpō to suitable and safe, predator-free islands. Effectively, this was the end of truly wild Kākāpō; all populations of the species would be, to a large extent, "managed" from now on.[43]

And so we return to Little Barrier Island, one of the most important places in the story of New Zealand conservation but for Kākāpō in particular. The first Kākāpō were transferred to Little Barrier from Stewart in 1982—nine females and thirteen males—and so too was Ralph Powlesland, more or less. Ralph was one of the small team of scientists who had painstakingly studied the newly discovered Kākāpō on Stewart Island and had been involved in every aspect of the subsequent debates and decisions about the bird's future, including the huge call to move the lot to other islands.[44] Although three other islands were used, Little Barrier was by far the largest and least disturbed. It was, however, much farther away—fully 1300 kilometers (807 miles) to the north—and would therefore provide a rather different climate and habitat compared to the bird's original home of Stewart Island. It was imperative that the translocated birds were followed carefully to assess how they were coping in their new home. To do so, all the birds were fitted with tiny radio transmitters and their movements monitored intensively for the first few years. Subsequently, tracking was less intensive and less invasive, involving indirect surveys using trained dogs.

To everyone's enormous relief, Kākāpō were found to be remarkably resilient, coping with the trauma of being captured, transported in boxes, and then released into an entirely new world. "Stoic" is the word Ralph uses. "They survived the drama of the transfer and nearly all went on to live long, normal lives on Little Barrier." The dedicated scientists tracking the birds found that some males soon formed their own "track and bowl" structures and that courtship booming was heard at several sites by the second year. Over the years that followed, almost all the males present were heard calling during most years; courtship by one sex at least was definitely under way. The problem was that the females did not seem to responding. Despite all that booming, there was absolutely no evidence of breeding for the first 7 years. Not even mating (as evidenced by a characteristic circular pattern of scattered feathers near the booming bowls) let alone nests, eggs, or nestlings.[45] This was a frustrating situation for Ralph and his colleagues, who were now spending half their lives—2 to 3 weeks at a time—on Little Barrier Island with their charges, enduring the rain, cold, and isolation. "The problem now was not survival," recalls Ralph. "It was them getting around to reproducing."

As one of the people to have made the link between the masting of podocarps and the breeding of Kākāpō on Stewart Island, Ralph soon realized that enhancing the protein content of the Little Barrier birds may be required. Neither of the two Stewart Island podocarp trees were present on Little Barrier, and although another tree species—the Kauri—producing potentially suitable seeds did occur there, its supply of seed was far more limited. Supplementary feeding with protein-rich foods was obviously worth trying, although just what to use was less clear. Furthermore, this was not a regular small bird that might visit a typical seed feeder; just how do you provision a large, nocturnal, ground-dwelling parrot that has apparently always perceived "food" as something that looks like vegetation? And another thing: it's often wet and there were rats (in this case, Kiore, the only remaining mammal on the island), which are likely to steal the food as well as being potential nest predators that you really don't want to benefit.

These were the formidable challenges facing Ralph Powlesland and Brian Lloyd as they began to plan one of the most unusual supplementary feeding experiments ever taken. How this was carried out is described in various scientific articles.[46] I had read these carefully many times but knew that there was much more to this remarkable story than would have been suitable for academic journals. To fill in the details, I was privileged to spend a few days with Ralph in his remote home in the extraordinarily beautiful Marlborough Sounds at the ragged top of the South Island (and within sight of Maud Island, once home to a bunch of lonely bachelor Kākāpō). Over many a coffee and dry Marlborough sauvignon blanc we discussed science, birds, and rugby, as well as supplementary feeding experiments.

The first problem was deciding where to place the feeders in the extremely dense forest landscape of Little Barrier Island when your target bird is effectively invisible. To find out where Kākāpō were likely to occur, Ralph, Brian, and their small team made hundreds of kebabs (or skewers) consisting of pieces of carrot, apple, and sweet potato (kumara) impaled on wire stakes and placed these at regular intervals along the network of tracks that had been formed to allow easier movements around the rugged terrain.[47] In places where the baits showed signs of having been "chewed," new stakes were moved progressively closer over several nights into a relatively flat site where the more substantial feeding station could be installed. Initially, the food items were simply placed in a dish on the ground, but the effects of weather and persistent raids by kiore forced progressive changes to the design of the feeder.

First, a circular tray was attached to a metal rod and positioned 350 millimeters (14 inches) above the ground, with a plastic cover above to protect the food from rain. Although the feeding apparatus became increasingly complex in design, the Kākāpō continued to visit and feed. But the device was about to get even more complicated. Rats were strongly attracted to this new foraging resource and, with squirrel-like tenacity, were soon finding ways to climb or jump up to the feeding tray. Even more serious was the effect of the weather. The combination of persistent wind and rain and, in the higher sites, heavy mists, was spoiling the food often within a single night. To make matters even worse, replacing the food regularly—sometimes daily—was proving extremely difficult; simply keeping the supplies available in this demanding landscape was taking up almost all the scientists' time. What was needed was a feeder that could protect the food from weather and rats and in sufficient quantity to last several nights yet still allow the Kākāpō to get at the food.

These were the sorts of fundamental engineering challenges that Ralph had grown up with on the family pig farm where he had seen his father's natural ingenuity at work, designing and constructing rat- and bird-proof food hoppers. Ralph's father was successful in thwarting all sorts of pests attempting to steal the pigs' food. Ralph and Brian's challenge was almost the opposite: allowing the birds in while keeping the damp and vermin out. The simplest solution was obvious but risky: a tight-fitting lid that covered the food but with a hinged flap that could be raised by a visiting Kākāpō. Although the birds had become familiar with the feeder and knew it provided food, the task of lifting the flap represented a major behavioral challenge. Plenty of species would simply be unable to make the connection between lifting the flap and accessing the food. But Kākāpō are not your average birds. After leaving the flap propped slightly open to allow the birds to get used to the construction, the flap was then lowered, requiring the bird to lift it with its head to access the food. Despite these challenges, almost all were feeding freely from the complicated feeding apparatus within a few days. The final innovation was a hopper made from plastic drainpipe with a liftable flat at the base that could be stocked with several days' supplies.[48] These hopper feeders were used for the more remote locations to minimize the amount of time spent visiting to add food.

Having solved the problem of feeder design, it was time to test what food types the Kākāpō would eat. The scientists started with a blank slate:

just about anything vegetable that could be obtained was on the menu. In all, 45 different foods of an extraordinary variety were offered. The feeding stations were visited each day and the remnants examined closely to assess the culinary preferences of the pampered clientele.[49] Only two items—dried figs and parsley leaves—were rejected entirely, while a further 20 were nibbled but not consumed, including banana, bread, and peanut butter. However, about half the smorgasbord of offerings was eaten regularly and these tended to be root vegetables, seeds, and nuts. Intriguingly, Kākāpō loved frozen but not raw corn on the cob. They also rejected raw potato yet devoured it if it was boiled or baked. These initial taste tests led to a short list of six foods that were both preferred by the birds and could be easily procured and stored for long periods of time: apples, sweet potatoes, almonds, brazil nuts, walnuts, and sunflower seeds. On the other hand, the birds may have loved the special porridge (oatmeal) the scientists ate for breakfast each morning but hospitality can only stretch so far.

It did take a while for a Kākāpō to establish a routine, but within a few months of the construction of feeding stations, almost all the birds were visiting their local canteen each night. And the birds were certainly consuming plenty of the supplementary food now available: the feeders needed continuous renewal and the birds' weight (recorded on an electronic scale hidden under the platform a Kākāpō stood on while feeding) began to increase almost immediately. Over the first 2 years of the initial trial, the average weight of the males went up by 42% and females by 36%.[50] But was there any effect on breeding activities, the main goal of all this work?

Although male Kākāpō on Little Barrier Island had engaged in their booming courtship displays most years since their arrival, in the first summer following the start of supplementary feeding Ralph and his colleagues noted booming every night and eventually determined that they persisted for much longer than any previous summer. In the second year, these boomers were at it nightly for 6 full months (October to April). It seemed that the additional food was stimulating the males, but what of the females? The first positive evidence that something may have been happening was the discovery of the telltale "feather clusters," places where contour and down feathers had been trampled into the leaf litter—clear evidence of vigorous Kākāpō copulation—only 4 months after the start of the experiment. Even more exciting was finding the first nests a few months later, located by following the now stationary radio tag to the carefully hidden incubation bird within dense undergrowth, in a hollow log or among the roots of a tree. Of the five radio-tagged females being studied, all of whom were utilizing the feeders, two produced eggs during the first year and four did so by the second year. The fifth female, labeled opaquely simply as "D" in the scientific articles but known as "Bella-Rose" to the scientists ("All of the Kākāpō had names," declared Ralph. "Every single one."), failed to show any interest in breeding though she did exhibit perhaps too much interest in her feeder. Bella-Rose features frequently in the supplementary feeding literature as a cautionary example: those individuals for whom the benefits of additional food can become a serious liability. Within the relatively short period of about 20 months she came to weigh in at 2.1 kilograms (4.6 pounds), 60 % heavier than the average of the other females and the heaviest female Kākāpō ever recorded.[51] That is quite an achievement for a largely "veggies and nuts" diet but given the intentional high protein composition of the food, not entirely surprising. Nonetheless, when a prominent outcome of an intervention aimed at improving reproduction actually results in obesity and a complete lack of breeding activity instead, concern is certainly justified.

Unfortunately, fat females were only part of the story. After the joy of finding nesting birds so quickly after the start of the feeding experiment, when it appeared that the addition of food was clearly having the hoped-for effect on breeding, both females abandoned their eggs for unknown reasons. The following year, with four birds nesting, only a single fledgling was successfully raised; a remote camera captured the juvenile feeding from a feeding station with its mother about 4 months after leaving the nest. Supplementary feeding seemed to be increasing breeding activities—booming and mating and even egg laying—but this was not necessarily resulting in regular and successful reproduction. The conclusion drawn by Ralph and Brian from this first experiment was that while the additional foods were obviously important and influential for breeding in Kākāpō, there was still a lot to be learned.[52]

Nonetheless, this exploratory study was an important first step in the long and laborious journey toward securing the future of this remarkable bird. Despite the many unknowns associated with the diet provided, there was no question that providing additional food was essential to the ongoing management of Kākāpō on islands such as Little Barrier. The feeding continued—using Ralph's special feeder designs—but understanding in more detail the physiological and nutritional requirements of the birds became a major goal for the next phase of the research. For example, years of careful attention to the bird's foraging activities, before and after the advent of supplementary feeding, indicated that the periodic masting events remained central to the control of breeding activity among Kākāpō. Even on Little Barrier Island, where only one mass seeding tree, the Kauri, was present, the birds appeared more likely to at least begin breeding activities during years when lots of seed was produced. Indeed, although the birds utilized their personal feeders extensively, the feeders provided only part of their overall diet. Apart from the unfortunate example of Bella-Rose, who appeared to forsake the chore of foraging on natural foods, most fed Kākāpō continued to browse extensively on the naturally occurring stuff. The scientists were pleased with this observation; they were initially concerned that making it so easy to find food may lead to a reliance on the feeding stations, but such fears proved unfounded. The Kākāpō still spent much of their night searching for the natural foods they subsisted upon.

Kākāpō remained terrifyingly close to the abyss of extinction so any assistance, including supplying additional food, was considered an important part of their conservation management. As the supplementary feeding continued, so too did efforts to understand the influence of food on reproduction. Over the long term, the key benefit of the supplementary foods seemed to be enhancing the physical condition of females so that they were better prepared for egg production in years when breeding might occur. Providing additional food supplies, however, did not influence the likelihood of breeding; that trigger seemed to be associated with natural cues, including masting. Nonetheless, food supplements seem to have resulted in more eggs being produced and increasing the survival of chicks when natural food supplies began to wane later during the breeding season.[53]

Supplementary feeding Kākāpō also had an important effect that no one saw coming. When the sex ratio of the offspring of fed birds was compared with those who had not received additional food, it was found that feeding was leading to far more male than female young.[54] And not just a slight increase: feeding seemed to result in two-thirds of chicks being male, while unfed females produced one-third. Both of the proportions are bizarre: in almost all species the typical sex ratio is remarkably close to 50:50. However, even among adult wild Kākāpō, the ratio was 62:38 in favor of males. The most obvious explanation is that females, engaged in prolonged ground nesting, are more likely to be vulnerable to predation (though booming all night from the same spot would surely be risky for males). Another possibility is that the production of more males is normal in Kākāpō, a view supported by remarkably detailed examination of ancient Kākāpō bones, which found twice as many males as females.[55] This idea actually accords well with a theory that mothers of some species may be able to manipulate the sex of their young.[56] The ability to influence the gender of offspring appears to be important in species where there are large differences in the reproductive success among adult males, which is certainly the case in Kākāpō. While all the adult males compete for the available females, few are chosen, but the chance of raising a stud is definitely worth the attempt when food is readily available, from an evolutionary perspective. Because young male Kākāpō are larger and grow more rapidly than females, they are more costly to produce because additional effort is required to rear them: producing males is simply more "expensive" than females. During regular years, females are "cheaper" but in good years, when females are in better physiological condition, producing males is not so costly. Providing additional food to females seems to have made any year a good year, with more males being the result.

This remarkable and unexpected finding has been used in even further fine-tuning of the supplementary feeding regime being used for Kākāpō.[57] While the objective of supplying additional food would appear to be all about the production of more Kākāpō in general, in the case of this weird species, more males is not going to help. The challenge is to feed adult females in such a way as to ensure that the nutritional threshold required for breeding is met (though apparently triggered by other factors), yet somehow manage to keep their condition low enough to avoid male-biased clutches. But wait: "A surer way of obtaining female progeny," explain the researchers, "would be to delay provisioning until after the eggs are laid, but this might [result in] lower egg production and poorer hatching success of eggs due to 'unfed' females spending more time foraging [and risking their eggs getting chilled]."[58] With both clarity and understatement, they conclude that this degree of micromanaging the supplementary diet: "clearly presents a considerable challenge for the Kākāpō conservation program."

Getting the Timing Right

I have spent a lot of time on these strange birds from a distant land—so vividly unlike the typical birds visiting our feeders—because these conservation stories center on the use of food as a key management tool. Working with species on the very edge of extinction, the extraordinary people involved in these programs have been forced to try all sorts of new approaches and solutions, whatever might work, to ensure that these species remain with us. With every egg and offspring vital, the level of scrutiny directed at individual birds in these species is extremely unusual, but the insights have often been invaluable and, in some cases, entirely unexpected. These findings lead to better understanding of the relationship between the birds' diet and reproduction—among other things—and on to ever more fine-tuning of the foods being supplied. For example, detailed analysis of the natural diet of the Kākāpō has led to the development of pelletized food made of nineteen different ingredients, resulting in larger clutches, a major win.[59] More recent iterations have involved targeting the best time of the year to provide the food.

Although the degree of intervention occurring in the New Zealand examples is somewhat extreme, it is by no means unusual in the hands-on end of active conservation biology. There are numerous examples of situations worldwide where the survival of extremely threatened species is based directly on supplementary foods. Spanish Imperial Eagles in Spain, Seychelles Magpie-Robins, and Red-crowned Cranes in Japan, to pick some far-flung examples, all currently benefit from carefully managed supplementary feeding programs. On the island of Mauritius in the Indian Ocean no fewer than four endemic species (Mauritius Kestrel, Mauritius Parakeet, Mauritius Fody, and Pink Pigeon) are effectively being preserved through supplementary feeding. In the case of the Pink Pigeon, whose entire wild population in 1990 was 10, the provision of additional food (just plain wheat) has raised the free-flying population to about 350 birds in 2006.[60] Interestingly, while almost all (99%) wild Pink Pigeons utilized the feeding stations at least once every few days, visits grew less frequent with age. Indeed, even in this precarious situation, over three quarters of the bird's feeding time was spent foraging on natural food plants. The importance of the supplementary supplies was in meeting the shortfall following cyclones and other periods when natural food may be difficult to find.

One (of the many) patterns emerging from such studies as these is that access to supplementary foods can be of critical importance at specific times in a species' annual schedule. These times may include the period leading up to the breeding season or when eggs are being formed or during the females' recovery time immediately after laying their eggs. Being able to feed on high-quality foods during these times has been shown very clearly to improve the adults' condition, providing a sound foundation for vital activities such as fighting for a decent territory, developing large, healthy embryos, and recovering quickly enough from the efforts of egg production to be able to assist in incubation and the raising of the young. The importance of food supply in these phases has been demonstrated in numerous supplementary feeding studies (as discussed in Chapter 5).

There is another phase of the breeding cycle in birds that is probably even more important to successful reproduction: the nestling stage. This period between hatching and independence, when the juveniles fly off into the sunset, is fraught with dangers and risks. Newly hatched baby birds—defenseless, vulnerable, noisy, smelly, unaware—are obvious targets for predators and highly susceptible to cold, wet, and wind. It is no surprise at all that this is the period when the greatest mortality occurs in birds.[61] Nestlings are also extremely prone to starvation if a sufficient and suitable supply of baby food is not available. Baby birds grow extremely rapidly, often doubling their body weight in a matter of days, a process entirely dependent on the ability of the parents to find enough food. In the vast majority of species, this means insects—typically larvae such as caterpillars and grubs—which the adult birds must find and bring back to the ravenous brood in an endless and exhausting process.[62] If the supply of bugs is insufficient, the effects back in the nest may be catastrophic. Competition among the clamoring chicks can become extreme, and the adults may even begin to deliberately favor certain individual young over others. The inevitable result is that some of or, all too often, the entire brood can simply die of starvation.

The link between reproductive success and the availability of insect foods is very well known and has been studied in a large number of bird species.[63] This has led to growing concerns over the possible effects of chemical pollution (including emissions from vehicles) and pesticide use in agriculture on insect populations in general, factors that have been implicated in the dramatic decline in many birds, especially in rural areas.[64] Whether supplementary feeding could play a part in this story has, however, not been investigated to any great extent. Almost all such experiments have involved providing various seeds and nuts and almost always avoiding the nestling period. There remains a strong (if probably misplaced) concern about nestlings being fed something—and possibly choking on—that is sensibly regarded as unsuitable food for nestlings. The obvious solution would appear to be providing insect foods instead, but this has been tried only a few times. One rare example of the use of insect food as a nestling period supplement is also important to our present exploration of the use of feeding for conservation. But this does not take us to some tropical island in search of an exotic species, but to a very different setting indeed.

Saving Sparrows

I am walking briskly through the wintry streets of East Dulwich, just south of the River Thames in inner suburban London. The sky is low and leaden and it is damp and chilly—about what you might expect for early winter in England—but the weather has not dampened the enthusiasm of my friend and guide, Dave Clark. For Dave, a visitor on a weekend is all the excuse he needs to scour his local patch for winter stragglers, birds that are late in leaving or those on their way south from Scotland or northern Europe. As we pass through numerous neatly maintained parks, unkempt vacant woodlands, communal garden plots (allotments), and large recreation reserves, I am amazed at just how much green space—and hence bird habitat—actually exists in one of the most densely urbanized, people-packed parts of the planet. "A lot of it is not that pretty, but the birds don't seem to mind," explains Dave as we stride through yet another park. "If you take the time and know where to look, it's amazing what you can find." As if to prove his point, a wild, wind-tossed group of Redwings rises from the lawns where they have been foraging on fallen berries. On a dull day like this I had not expected that we would see much but, for me, the sight of these gorgeous thrushes—along with (among others) a Goldcrest, Mistle Thrush, four species of tit, and even a kestrel sheltering in the crumbling steeple of a neglected church—has quickly dispelled my notion of a big city being largely devoid of birdlife.

Later that afternoon as we reviewed the day's wanderings over a welcome pint in a local pub, Dave posed an arresting challenge. "No doubt, you will have noticed what we *didn't* see today." When I failed to respond (I had considered saying, "The sun!"), his answer was provocative and even a little alarming: "House Sparrows!" We had spent a good few hours wandering the streets, parks, and back lanes of London yet had not detected a single "Spadger" (or *Spuggie, Spurdie, Spurgie, Sproug, Speug, Lum Lintie, Craff, Cuddyrowdow, Thack,* or *Thatch,* as this once ubiquitous urban denizen has been variously known).[65] Once I had considered it for a moment, the significance of this observation slowly grew. There may be no greater example of a bird being synonymous with a city anywhere than sparrows and London.[66] The English capital was a key source of enormous numbers of sparrows exported throughout the world at the height of the British Empire's global expansion. Homesick Englishmen arranged for thousands of the hitherto disparaged little birds to be shipped out to the farthest reaches of the empire as an attempt to install a modest reminder of home. It was an enterprise extraordinarily successful: within a remarkably brief time (mainly the first half of the nineteenth century), the heart-warming chirping of London spadgers was heard from New York to Buenos Aires, from Harare and Cape Town to Adelaide and Dunedin.[67] (And even the remoter towns of Australia. Having been liberated at various places starting in the 1860s, House Sparrows—birds typical of the damp English landscape—were the most abundant species I counted during the 1980s in both the hot, dry inland town of Wagga Wagga as well as the tropical streets of Townsville in northern Queensland.)[68]

My, how times have changed for the sparrow! Once among the most abundant bird species on the planet with one of the greatest global distributions, House Sparrow populations are now in precipitous decline throughout its range.[69] While there are few places where the birds have disappeared entirely, their numbers are falling almost everywhere. In many of these places, this is of rather little concern as sparrows are often regarded as invasive intruders, frequently accused of stealing grain intended for farm animals or people, damaging crops, and building unsightly, fire-prone nests in buildings.[70] No, in places where they were introduced, their apparent demise has largely been welcomed, if it has been noticed at all.

But the story has an entirely different complexion in the House Sparrow's natural distribution. Throughout much of continental Europe and the British Isles, the species appears to be disappearing rapidly, although the pattern is extremely uneven.[71] For example, European House Sparrow numbers have declined markedly in Ghent and Hamburg but much less so in Berlin and Paris.[72] Similarly, in the UK, where the most detailed analyses have been undertaken, the abundance varies enormously across the landscape, and although the number of sparrows associated with farms has declined considerably, these sites continue to support the highest densities.[73]

The most alarming declines, however, are very clearly associated with suburban landscapes that, in the UK, represent among the most extensive and important habitats for a wide variety of birds. Within Britain's towns and cities, sparrow abundance is typically highest in private gardens and allotments; not unexpectedly, therefore, the most dramatic losses in sparrows have been in areas where these habitats have been lost to, typically, housing and parking lots.[74] This is especially serious because these places are increasingly targeted for "urban infilling" as more people are being crammed into whatever "vacant" space remains.[75] But this generalized loss of space and habitat is definitely only part of the sad sparrow story. House Sparrows, like many Londoners, seem to be able to cope more or less happily with high-density living. While their abundance in an area may have been drastically reduced, groups of these birds can hold on in even tiny patches provided a few basics are available: a nice dry niche under the awnings for nesting, some dense bushes for shelter, and perhaps access to a chicken shed, stable, or even a bird table for some seed. Across the landscape, there are thousands of such refuges. The problem is that even these resources are becoming harder to find. Farmyards and back gardens are becoming increasingly tidier, with less uncut grass (reducing the supply of seeds) and more secure holdings for farm animals, while newer house designs are reducing the traditional nesting places such as the openings under eaves and roof tiles.[76] The widespread practice of replacing old wooden soffit boards with much more tightly fitting plastic versions, for instance, is quickly reducing the availability of an important traditional nesting place. But even where all these factors are present, many populations in urban areas in the UK continue to decline. Overall, the British population of House Sparrows is estimated to have crashed by almost 70% over only a 26-year period (1977–2003), a terrifyingly rapid fall.[77] Once the quintessential English bird, common everywhere, the species finds itself listed on the Red List of Threatened Species.

This is a concern that Will Peach of the Royal Society for the Protection of Birds (RSPB) has taken to heart. I met with Will at the RSPB headquarters north of London (described in Chapter 3), and while we discussed a lot of topics, our main focus was on Will's extensive sparrow work. As Will acknowledged, "[Sparrows] have dominated my life for more time than I would like to say." One of the first studies was an attempt to understand the condition of the sparrow population in Leicester, a large industrial city in central England.[78] This very detailed work, undertaken mainly by Kate Vincent, made two important discoveries: an unexpectedly large proportion of sparrow hatchlings died during their first week of life, and the weight of chicks was much lower than it should have been just prior to fledging. Together, these key parameters seem to have led to a level of reproduction simply too low to maintain the population. Although various factors may have been implicated in this outcome, analyses pointed most clearly to an insufficient supply of insects in the diet, particularly in the first few days after hatching. This was evident in both the inadequate levels of invertebrates (especially aphids) and atypically high levels of vegetable material (as found in the droppings of nestlings). Baby sparrows were being fed too much plant stuff (mainly bread, peanuts, and seeds, almost certainly coming from feeders) and not enough good insect protein, presumably because the latter was simply hard to find. We can be sure that the parent birds know what is needed; it's just that these were circumstances of desperation.

An obvious scientific approach would be to see what might happen if insect food was provided, particularly during the critical period when the newly hatched chicks are growing rapidly. Collecting the naturally occurring insects most commonly used by sparrows, however, would be just a little too challenging. (Just how many aphids would be needed each day?) Thankfully, an insect food is available commercially: mealworms, which are actually the larvae of a beetle, *Tenebrio molitar*. These hardy and increasingly utilized additions to feeder offerings were the clear option for an important supplementary feeding experiment. And the spectacular loss of sparrows from the suburbs of London (a drop of 85% from Kensington Gardens between 1925 and 1995, for example) made the capital an obvious place to undertake such a project.

The study that was carried out remains one of very few supplementary feeding experiments conducted in an urban environment.[79] Because it was important to see what happens in the typical places where sparrows live, Will and his colleagues recruited suburban Londoners with private gardens already supporting sparrows, who were willing to participate in what was quite a lengthy project (lasting for 4 summers). Specifically, the goal of the study was to determine whether providing insect food would alter reproductive success and, ultimately, the number of adult sparrows. To address these aims, suitable sites (private homes with gardens) were divided into two groups: those where the sparrows were fed mealworms and those where the local sparrows got nothing beyond existing feeders. For the "feeding" sites, volunteers were supplied with sufficient supplies of mealworms to allow both morning and afternoon offerings throughout a 16-week period over the summer. About 100 g (3.5 ounces) of mealworms were provided daily, amounting to 11.5 kilograms (25 pounds) of live grubs for the entire season (about 100,800 individual mealworms). While the residents were topping up their feeders, Will and his sharp-eyed assistants made careful, unobtrusive (not everyone was likely to believe the explanation that they were peering intently into private gardens to look for sparrows) observations of "chirping" male sparrows producing their familiar territorial call, females, and especially the telltale fluffy fledglings. In these authentic conditions, as opposed to the usual experimental studies using easily accessed nest boxes, finding and examining sparrow nests was virtually impossible; old-fashioned observations would have to do.

By the end of this project over a ton of mealworms had been supplied—and eaten quickly—by the presumably grateful sparrows of London. So what effect did this vast amount of additional animal protein have on these populations? Thankfully, there was an important result, at least in terms of the numbers of fledglings being produced. Specifically, all those mealworms meant that a higher proportion of eggs hatched successfully and led to 62% more fledglings compared to the unfortunate sparrow colonies that missed out on the food.[80]

At the halfway point in the experiment, 2 years in, a review of progress confirmed that the addition of the mealworms was definitely having the hoped-for influence on the first aim—enhancing the breeding success of the sparrows—but strangely this was not translating to larger colony size. The birds were raising more chicks but not adding more adults to the population. The second and crucial aim of increasing overall sparrow numbers was not going according to plan. Will and his team decided, therefore, to turn their attention to providing a supplementary food that should benefit the adults too: rich, high-energy sunflower hearts. So, in addition to the summer mealworms, for the second 2 years of the study the researchers provided enough bags of sunflower hearts to the willing London residents so that the provisioning could be continuous: energy-rich food, on tap, every day, for 2 whole years.[81] A metric tonne (over 2204 pounds) of mealworms and 7.5 tonnes (165,000 pounds) of sunflowers over the full 4 years were supplied: now that is what you can call a serious supplementary feeding experiment.

This major study is particularly important because it closely resembles what is happening in the typical suburban environment in which a large proportion of wild bird feeding occurs. Because of its scale, all our private feeding can be regarded as a truly gigantic feeding experiment. And because of its clever design and duration, Will Peach's London experiment may be about as close as we can get to discovering what might actually be happening in the suburban wilderness. I was, therefore, more than a little keen to hear the outcome. Did the provisioning save the sparrows?

"We must have finished the London field work over 5 years ago, I think," recalled Will, the substantial effort involved clearly an uneasy memory. "All those containers of hand-delivered mealworms! All those bags of sunflowers! The logistics, the travel, the time, the people; it was a big exercise, you know." He pauses to sip on his now definitely cold coffee. "When we finally cranked the data—and that was a challenge itself, trying to account for so many variables—we had two clear findings. First, the mealworms definitely had had a large and positive impact on the reproduction of the sparrows being fed. There were certainly more eggs hatching and chicks fledging. And so, when we added continuous supplies of high-quality seed, did the abundances go up? Were we actually producing more adult sparrows? Was that huge effort worthwhile? Well, shockingly, no. Not really. Overall, our 4 years of feeding increased the abundance of territorial male sparrows by a bit—about 8%—and the effect of feeding was only statistically significant in the small colonies." I, too, was astonished. There was obviously more to the decline of the sparrow than simply food supply, though the clear impact of providing plenty of insect food was a very important insight. "And it was not just a London issue either," Will continued. "We repeated much the same experiment back in Leicester,[82] where we knew the situation pretty well, and got almost identical outcomes: healthy chicks and an increase in breeding success but that's about all."

Why?

Will's hard-won conclusions: the supply of insect food is indeed critical but it does not drive the population; the availability of feeder foods is only part of the picture, at least with House Sparrows and probably other smaller suburban species that make use of our feeders.

"The Most Fantastic Bird Table in the World"

While London sparrows have plenty of challenges, one thing they have not had to worry about (OK, it's the people who are doing the worrying) for a very long time was being eaten by a bird of prey. For centuries, large raptors were abundant and conspicuous throughout the city; Shakespeare even described London as "the city of kites and crows."[83] One species in particular, the Red Kite, was very well known, thriving on the easy meals provided by the haphazard disposal of organic waste by Londoners. (Numerous travelers long ago had noted that one always knew when they were approaching London by the increasing stench.)[84] Despite the public service Red Kites carried out, of reducing the amount of rubbish rotting in the streets, even this well-adapted urban bird of prey suffered from the widespread antipathy toward predators generally. From royal decree to gamekeeper's disdain, British birds of prey of all species—as elsewhere around the world—were subject to unrelenting persecution wherever they occurred. By the 1870s the Red Kite, though formerly a familiar sight, was extinct in England.[85] At the start of the twentieth century, the entire British Red Kite population had been reduced to a handful of birds that had somehow withstood the shooting, poisoning, and nest destruction, hidden away in the remote hills of mid-Wales. Perversely, this very rarity posed new threats as egg collectors sought what had become a highly prized item. The future of the Red Kite in Britain appeared extremely bleak indeed.

The fortunes of the British Red Kite began to change at least marginally with the formation of the first Kite Committee in 1903, a group of dedicated amateur naturalists deeply concerned about the plight of the species.[86] Their first actions aimed to protect the nests from egg collectors, who were destroying about a quarter of all clutches at the time. Their efforts were quickly acknowledged by the recently formed Royal Society for the Protection of Birds, which, in 1905, became involved in what eventually became a flagship campaign for the society. The extraordinary resolve of the RSPB and its various partners has led to the longest continuous conservation project in the world. Today, we can also add that, as well as longevity, the Red Kite campaign has also been one of the most successful—though this was not evident until relatively recently. For much of the early 1900s, progress was slow and the persecution relentless. By the 1970s, however, the egg collecting fad had largely petered out and the indiscriminate shooting of any bird of prey was far less common. Nonetheless, the miniscule size of the remnant kite population meant that natural growth and dispersal were virtually nonexistent. Without radical intervention, local extinction remained a distinct possibility.

The turning point came in 1989 when RSPB and the English Nature Conservancy Council (now Natural England), following the strict guidelines of the International Union for the Conservation of Nature, released the first Red Kites seen in England for over a century.[87] (A similar exercise was also undertaken in Scotland at about the same time.) The site selected after painstaking research was a reserve classed as an "Area of Outstanding Natural Beauty" in the Chiltern Hills about 35 kilometers from central London and 20 kilometers north of Reading. The aesthetic appeal of the chalk escarpment of this area was about to be significantly enhanced. Because the Welsh population of kites was far too small and vulnerable to provide a source of new birds, the dramatic decision was made to import kites from the healthy stocks of Spain. Over a period of 5 years, ninety Spanish Red Kites were brought to the reserve and, after a prolonged period of settling in on-site aviaries, they were progressively released into the rolling terrain. Only 3 years later, in 1992, the first successful breeding was detected and, from that moment, the birds have literally taken off.

Today—only a few decades later—the local population is estimated at well above 500 and more than 1000 pairs throughout southern England.[88] In February 2006, for the first time in 150 years, Red Kites were again spotted in London—Hackney to be precise—with considerable excitement (though probably less positively among the local birds).[89]

The success of the Chilterns experiment quickly led to similar projects throughout the United Kingdom and Ireland, with birds sourced from Sweden and Germany as well as Spain, and, significantly, also from the now burgeoning Chiltern population itself.[90] The result: the remarkable return of one of Britain's avian icons. Red Kites have become a regular and spectacular sight, especially in the southeast of England. Along the M40 these days, it's hard to miss them. And, of course, this dramatic recovery has been based squarely on supplementary feeding. As we will see, however, this can occur in official as well as less formal ways.

One of the features of Red Kites that has undoubtedly aided in their restoration has been their broad and nondiscriminatory diet: they seem to eat just about anything (in *King Lear*, Shakespeare has them stealing underwear from washing lines, but probably for nesting material rather than for food), though some sort of meat is preferred. They tend to be scavengers rather than hunters of live prey and throughout their distribution are known to feed primarily on roadkill and dead farm animals. While such items are certainly available in the rural landscapes of the Chilterns and the other places where they are now recovering, the prospects of the kites has been enhanced by the provision of varying amounts of readily accessible meat. The birds intended for release are imported as juveniles 4–6 weeks of age and spend a couple of months in their large, open-air aviaries, getting used to their new surroundings. During this time they are fed a varied diet consisting of abundant animals collected by local foresters: rabbits, Gray Squirrels, Fallow Deer, Muntjac (a small invasive deer) as well as Woodpigeons and crows.[91] These are just the foods the kites are likely to find in the local area, and they are offered in large chunks or whole, requiring the birds to tear the items apart themselves. When the young birds are released, some of these items continue to be supplied near or on the aviary but only for a few weeks. Although a few individuals return to the release site, most move away within the month.

Generally, this type of feeding should only be necessary during the early stages of the restoration of a release population. An obvious parameter of successful establishment is the ability of the birds to sustain themselves without being overly reliant on external food supplies. But of course, there may be other reasons for persisting with the feeding, especially when the results may be downright spectacular. The most famous example is the Gigrin Farm in central Wales.[92] In the late 1980s, the long-term owners, the Powell family, were supporting the few local Red Kites by supplying some rabbits shot for the purpose. This was a somewhat unusual practice for Welsh sheep farmers, but the Powells appreciated the dire state of the kite at the time and were willing to put aside the traditional antipredator stance of their colleagues. In 1992, the dedication of the Powells was recognized by the RSPB, who approached the family to see whether they might expand the scale of the feeding and become an official Red Kite feeding station. In addition, they asked whether they might also consider allowing the public onto the farm to view the process. This was an unusual request for a successful livestock enterprise, yet the Powells agreed. The exercise began modestly, but as the number of Red Kites began to rise, so did the number of paying visitors. The popularity of the spectacle has resulted in the still working farm becoming an internationally renowned attraction. From the original 3 pairs, today between 200 and 600 kites can be seen, attracting many thousands of people annually.[93]

At exactly the same time (2:00 p.m.) every afternoon of the year, piles of roughly chopped meat of various sorts (the local rabbit population could never supply the required amounts) is dumped in front of 5 giant hides providing a spectacular view of the ensuing melee. Red Kites maneuver brilliantly through the air, wheeling and spiraling with astonishing precision. Needless to say, the huge pile of food also attracts a range of other meat-eating birds—buzzards, Jackdaws, Ravens, and crows—though the imposing size and demeanor of the kites tend to dictate that they consume the bulk of the offering. The sight of hundreds of massive birds of prey interacting with each other and these other species is unquestionably remarkable and memorable. The BBC described it as "the largest, most spectacular bird table in the world!"[94] It is also an extremely successful tourist attraction. What is less clear is the possible impact of this massive and dependable source of food on the local ecology. There is no doubt that this long-term supplementary feeding exercise has greatly enhanced the recovery of what was an iconic species on the brink of extinction. But is the obviously unnaturally concentrated aggregation of kites a sign of conservation success or (to be provocative) a contrived artifact mainly for human "consumption"? The Powells of Gigrin Farm are certainly aware of these concerns and point out—accurately—that the kites spend most of their day, especially during the morning, hunting naturally over a vast area before turning up for feeding time at the farm. And the numbers of kites fluctuate markedly, with the highest numbers directly associated with prolonged poor weather when hunting is hard. The main role of the feeding at Gigrin is, according to the Powells, to provide "a top-up or emergency ration."[95]

Perhaps the knowledge that supplementary feeding has apparently assisted in bringing back the Red Kite is all the evidence that is needed. Indeed, when it is possible to see this wonderful bird in one's own suburban back yard, as is now the case in lots of places around the Chiltern Hills, for example, maybe even private householders can help. Certainly, this was a prominent finding of a survey conducted by Melanie Orros and Mark Fellowes.[96] As described in Chapter 4, these researchers heard that people living in villages and towns in the vicinity of Reading spontaneously started to feed the kites they began to see passing through. They suspected that maybe a handful were involved; their research revealed at least 4500—about 5% of the residents of Reading—happily admitted to feeding kites. Around half of these people stated that they did so "to conserve them," while an even higher proportion simply "wanted to see them close up." Providing meat as a feeder food is highly unusual in most suburban gardens, raising all sorts of concerns such as attracting vermin and encouraging predators of typical garden birds. Nonetheless, the possibility that people might feed Red Kites anyway prompted several organizations to produce advice and guidelines on the best way for householders to feed these birds.[97] These suggestions were quite specific and aimed to provide a diet suitable to a bird of prey: ensure that the items contained skin and bones (whole, small rodents were ideal); avoid cooked or processed meats; and don't use roadkill animals (which could be poisoned or in other ways toxic).

Few of the people feeding the kites, however, seemed to be aware of these guidelines. Many offered whatever food was readily available, typically using chicken and other convenient meat scraps.[98] Their engagement seemed to be based on the rather straightforward motivations associated with almost any feeding: to see the birds and hopefully help them. Their impulses were aided by the facts that this new "garden bird" was also a large, unusual, and impressive species, one widely known to be of great conservation concern. Indeed, given these features, it almost seems surprising that not more feeders were involved. (Although, come to think of it, they are birds of prey.)

Unsurprisingly, the issue of feeding Red Kites in suburban gardens has proved controversial. For some people, the increasing abundance of big raptors in town has led to genuine concern for the plight of their familiar little garden birds. Even seemingly well-informed writers have expressed unease at what has been described as the "artificial situation of kites benefiting from garden feeding," suggesting that this would probably lead to the survival of those individual kites that would normally die (being weaker or less capable) as well as increasing the numbers of these birds in the places where "human-provided foods" were available.[99] That these are common and even desired outcomes of almost all wild bird feeding appeared not to be appreciated.

The remarkable return of the Red Kite (so far at least), ironically, may have been almost too successful. Several people who had been directly involved in the development of the kite-feeding guidelines have since withdrawn their support and now actively oppose the practice.[100] As the species' abundance and distribution—especially in urban areas—has increased, so have negative comments and sensational reports in the media. There may be more than a hint of traditional antipredator bias in the numerous stories ("Kite Tried To Eat My Cat!," "Children [who were actually hand-feeding a kite] Severely Scratched by Wild Hawk!," and "Enough Is Enough: Time for a Cull!") that appear regularly. Intriguingly, a prominent ornithological organization now refuses to be associated with the feeding of Red Kites, not because of any scientific evidence, but because it does not want to be associated with a potentially negative conservation story. This is a sobering reminder that perceptions rather than reality often color our reactions. What does the typical householder see when the (still endangered) kite is seen eating a (common but ever so cute) Blue Tit?

Could a Feeding Station Be a Trap?

From a mere handful of birds in the whole of the UK in the 1980s, Red Kites now number over 3000, a remarkable recovery strongly associated with the provision of additional foods. An important measure of the success of this particular program is that the birds are only briefly reliant of this food; after only a few weeks they have moved away to foraging on a wide range of naturally occurring foods. Unexpectedly, these include items intentionally supplied by people in their gardens, though there is no suggestion that the kites are dependent on these sources.

This is not the case for certain especially vulnerable species where supplementary feeding is a key component of their conservation plan. The feeding regimes provided for the Takahē and Hihi on Tiritiri Matangi Island and other species with tiny populations are essential for the survival of the species involved. The withdrawal of these supplies would be very likely to have serious consequences for these species, primarily because of their extremely small population sizes and often the peculiarities of their diets. For other species being carefully managed for conservation, however, the objective is normally for the species to be able to persist without becoming dependent on the supplementary food supply. Places that support animals because of the availability of some key resource, especially food, may sometimes represent a misleading indication of its overall quality as a place to live. Food is essential, of course, but if that is the only feature used in selecting a breeding territory, for example, the animal may potentially be making an ill-formed decision. If what appears to be a wonderful food supply was to be interrupted or changed to something else, or the site attracted too much competition or predators (and so on), deciding to set up nearby may be a serious mistake.

The idea of animals being drawn to a particular location mainly because of a certain attraction, even though the spot may actually be of poor quality or even downright dangerous, is known formally as the "ecological trap" concept.[101] This important idea is of direct relevance to thinking about feeding stations, both for conservation as well as in gardens. All sorts of examples of this process have been discovered—Indigo Buntings in North America, for example, prefer to nest in the sharp transitions between forests and grasslands, but when this natural preference led them to nest in the artificial edges caused by human clearing, they become much more vulnerable to predators that operate along these zones.[102] Ecologists have found that the rapid changes associated with urbanization provide plenty of opportunities for such selections, leading animals to settle in apparently attractive places that turn out to be inhospitable.

The potential that supplementary feeding may be associated with ecological traps is a constant concern for conservation biologists attempting to assist threatened wildlife. When the natural unpredictability of foraging opportunities is replaced by a feeding station that never runs out or moves, it is hardly surprising that many animals set up home ranges nearby. If they then come to rely on this supply instead of foraging more widely, the consequences could be significant. For these reasons, many conservation feeding plans actively attempt to make the provisioning less predictable, moving the stations around, varying the timing of provision, and often having an eventual cessation of feeding as a goal. Obviously, the influence of these changes have to be monitored and managed carefully; if the alterations are too abrupt or unexpected, the impact on the target species could be serious. When the feeding stations provided for endangered Spanish Griffon Vultures were suddenly closed, for instance, the population stopped growing and attacks on local livestock began to escalate.[103]

There is much of relevance here for the practice of wild bird feeding. Could our gardens potentially be acting as an ecological trap for the birds visiting our feeders? If we stopped stocking the feeders, would they be affected? Should we also emulate the unpredictability of nature and be less predictable? Unfortunately, these are questions without clear answers at present. Much more research is required.

Meanwhile, we will continue to feed. It's time to ask perhaps the biggest question of all: Why?

8

Reasons Why We Feed Wild Birds

As we draw near the end of this journey into the intimate and personal yet thoroughly commercial and global world of people and the birds they feed, it is time for a little reflection. For me, this has been a long and haphazard path, without a distinct beginning and a far from certain conclusion. Who can say where or when a lifelong obsession really starts? Some of my key childhood memories seem to include animals being fed. I recall chaotic family picnics that often finished with leftovers being tossed to ducks and geese, barbecues in parkland with bits of burned sausages being offered to pushy kookaburras and overly friendly kangaroos. Fish and chips by the beach and the ever-present gulls, lunchtime breaks during long hikes when all manner of wildlife would appear at even the most remote locations. Offering a tidbit to these expectant visitors was just a normal part of our informal family policy. "Country people share," was how my mother explained it; the unstated comparison with city folk didn't need to be made explicitly. Country people were also polite.

Feeding birds at home, on the other hand, was different. "Can't have them getting used to handouts. Too much of that sort of thing these days," explained my father, apparently without contradiction. "Everyone has to earn their own living." Now that I think about it, maybe it was my eventual move to the city (to earn my own living, Dad) where this interest really began. I don't recall anyone with a feeder during my childhood (though I may simply not have noticed), but plenty of the city folks I eventually met certainly did. Even though it would be some time before I began to take a more serious interest in this practice, there were plenty of occasions when a visit with friends included mention of birds being fed. Birds can be great conversation starters.

I am writing these words while sitting on the veranda of a café in the tropical rainforests of northern Queensland, Australia. My companions have left me to work (or at least think) while they wander along the paths in the forest directly behind the café. I watch them disappear into an impossibly complex wall of lianas, ferns, palms, and other vibrant plants of every green imaginable. A few meters from where I sit, a long rectangular platform feeder is suspended from the branches of a sprawling tree. A remarkable mixture of items has just been emptied onto the tray: pieces of banana, pawpaw (papaya), grapes, apple, and kiwifruit as well as several piles of seed mix. In an instant, the platform is swirling with a dozen species of bird, not one of which I have ever seen at a feeder anywhere before. Bridled and Macleay's Honeyeaters, Bower's Shrike-Thrush, Figbirds, a Tooth-billed Bowerbird, and even a magnificent Victoria's Riflebird (a type of bird of paradise). It is a spectacular selection of the birds people travel vast distances to see in these rainforests—and here I am ticking them off between sips of locally grown organic coffee.

Obviously this type of bird feeding is all about arranging a wonderful display for the tourists, as occurs all around the world. Certainly the numerous international visitors are appreciative, although after a frenzy of phone photos they drift off to the next attraction. I soon find myself virtually alone again, though I am quickly distracted by the arrival of a solitary Emerald Dove to the feeder. These are normally shy birds, and I am quite amazed to be sitting so close to one as it casually picks at a grape. "This one has only just started visiting," whispers a female voice from behind me. Camilla quietly introduces herself as the owner of the café. She and her young family live in the lower level (the café is on the second floor, providing a better view of the forest behind), having moved here from Spain about six years ago. "The previous owners started the feeding but they only put out sunflowers. I was sure we could attract more birds and so did a little research. All these different foods work so well. Even some of the local people are amazed."

With no new customers around, Camilla is only too happy to chat. "Of course the tourists love to see the birds, but this is not why I feed them. At home in Seville, there was no nature. It was all concrete and cars. There were no birds but pigeons. I wanted somewhere my children could see that there are still beautiful natural places. When we bought this café, it was just a way to make money. I had no idea that all these birds were around. When I saw the feeder I just didn't know what to do with it; we had never fed birds in Spain. But the previous owner explained that many of the people who came to this café were here to see the birds. It was part of the business. So at first I just put out the seeds and didn't think much about it. But now it means so much more to me." Soon after her family had arrived, Camilla describes how her six-year-old daughter came rushing inside early one morning before the café was open. "'Mama, quickly! There is monster in the garden!'" Camilla tells me her daughter said. "I now know that it was a cassowary," she says, "but neither of us had ever seen a bird so huge and so beautiful. It seemed impossible that something so wild could be just there, in our yard. We both just sat down and watched for ages. I then noticed that it was picking up the spilled seeds from the feeder above. I very slowly went and got some grapes and tossed these out onto the lawn. The bird took no notice of us, gobbled the grapes, then just walked back into the forest. It was just amazing. That morning seemed to change everything. I started to learn more about the birds in this land."

Some months earlier, this region had experienced the effects of a major cyclone, with considerable disruption to settlements along the nearby coast. While the damage to buildings, bridges, and roads had been considerable, there was also massive destruction to great swaths of the rainforest. Although many trees had been blown over, what was much more serious was the amount of fruit that had been destroyed. The storm had hit the area at the peak of native fig production, decimating the natural food crops that many rainforest birds rely on. Big birds like cassowaries were particularly hard hit. Throughout the region, these normally secretive birds of the deep jungles began to venture out looking for food. In several communities, people got together to supply fruit for the starving birds, although even this helping hand proved controversial. Inevitably, as more and more cassowaries began to wander into the towns, there were several vehicle collisions and even dangerous interactions between birds looking for food and people, including tourists and others who did not realize what was going on. Pro- and antifeeding factions formed, sometimes between neighbors and within families. These tensions continue to this day. In the parking lot outside the café, bumper stickers declare: "A fed cassowary is a dead cassowary."

Camilla's café is some distance from the coast and, although the cyclone was a lot less destructive up here, the impact on the frugivorous species was clear. It was when rarely seen birds such as Rose-crowned Fruit-Doves—strict fig eaters—began to turn up at the feeder to peck at the sunflowers that Camilla decided that her menu needed major revision. The response of the birds to the colorful smorgasbord of fruit, some grown in the garden below, was immediate; all sorts of species began to partake. "But we do have some rules too," she explains. "The tourists like to see the birds. Of course. But the feeding is really for the birds. If the weather has been kind, I put out just a little food, and only three times a day. It is soon gone and the birds have to go back to the forest where they can find their own food. After big storms, though, I know they are hungry, so they will get a little more." Throughout this lively exchange I said rather little. Once Camilla realized that I was genuinely interested in her story—and that I seemed unlikely to be critical of her feeding ("Some people can be very rude about what I am doing," she says)—she was only too willing to tell me all about her experiences as a bird feeder. "This is really important to me," she stated earnestly at one point.

The café's door chimes announce the arrival of more customers. Camilla disappears inside leaving me to consider what she has been saying, the various dimensions of her narrative. There is a lot to consider: cultural adjustments, a willingness to learn, the wonder of new discoveries, recognizing need, and then responding appropriately, even coping with criticism. But shining through it all was the expression of enthusiastic caring. I am certain that Camilla would dismiss my academic dissections with a vibrant laugh, saying something like: "Oh, you scientists just make things complicated." But while her remarkable story can hardly be regarded as typical of most garden bird feeding, I am more convinced than ever that almost all of us engage in feeding for a complicated web of reasons, even though we might simply say: "I just like birds."

Letting People Tell Their Story

As with so many other aspects of this journey, an initial attempt to tease apart the motivations people have for feeding wild birds started some time ago with a particularly insightful student. Peter Howard was an experienced social scientist who became interested in trying to understand what people thought about a range of human-wildlife interactions and conflicts. These included home invasions by animals such as possums and snakes or attitudes to feral dogs and foxes in urban areas. These interactions tended to be fairly negative, so Peter was pleased to include bird feeding as something much more positive. While I might know how to catch a kangaroo or radio track a koala, Peter was skilled in what I regard as an exceptionally difficult task: how to talk to people, anybody, and somehow get them to really open up. And not just talk; as Peter tried to explain patiently numerous times, it was important to get people to talk without them simply "saying what they expected them to say." As I have come to appreciate, it is all too easy to bias the answers of your participants simply by wording the question the wrong way. Asking someone whether they feed birds immediately after a question about with how feeding may harm birds, for example, is unlikely to lead to reliable answers. Peter's genius was in establishing settings where people felt relaxed and free to be honest and unhindered in their responses: "Just letting them tell their story," as Peter described it. This could be in a face-to-face interview, a chat over the phone or in the way questions in a survey are worded and constructed.

Peter's study (completed in the early 2000s) has been widely cited because, remarkably, this was the first time that anyone (as far as we could tell) had systematically asked people who fed birds to try and explain why they did so.[1] Although there had been a number of studies of the practices and scale of bird feeding by then, the motivations behind feeding were still unclear. Obviously, there were plenty of informal and anecdotal explanations around, but we wanted to know what the feeders themselves thought. In the absence of any prior findings, Peter initially spoke to a number of wildlife professionals with some experience on the topic to ascertain the general issues that needed to be canvassed. The next step was to speak to a selection of feeders directly. About 20 people were recruited through advertisements in the media, and Peter chatted at great length with these feeders. This was a crucial step because it soon became evident that many were extremely skeptical about our motives. As explained earlier, the clear although unofficial "official" stance in Australia is that all wildlife feeding is bad. This was something every feeder was very much aware of. Was this apparently friendly bloke who wanted to ask about feeding part of some tricky government trap? I am being perfectly serious. Although no private feeder had ever been prosecuted, the ubiquity and determination of the semiofficial antiwildlife feeding stance has plenty of people deeply concerned and suspicious.[2] Once Peter sensed this concern, his approach changed to one that emphasized that his aim was to assist feeders in caring for "their" birds. It took some time but this approach eventually worked, with many of these people becoming sufficiently convinced that they were able to encourage their friends to participate in the next stage of the project, a detailed mail survey. Over 150 people completed this questionnaire, and once the replies had been received, Peter and I hosted an informal gathering of about 40 of the participants where we presented our findings and, again, listened to the feeder's responses. This was an extraordinary evening to which we will return shortly.

Although the survey covered a number of familiar topics (food used, main concerns, favorite species, etc.), a key question asked feeders to describe, in their own words (a very important aspect of this type of research), their motivations for feeding wild birds; they could give as many reasons as they liked. A critical part of Peter's evaluation process involved carefully considering the responses in order to identify general categories for the reasons being expressed. After considerable deliberation, Peter decided that there were five main reasons overall, with most feeders mentioning at least two. Looking back at these results, I am reminded of how mundane and obvious some things were while others were entirely unexpected.

By far the most frequently cited reason for feeding birds—mentioned by over 75% of respondents—was along the lines of "It makes me happy."

That was hardly surprising, and made for a nice uncomplicated dominant category: "Pleasure." Other reasons were categorized as "Educative" (about 30%), with feeders indicating that they were able to learn about the birds when they visited, and "Attracting" (23%), where an important reason for feeding was to bring the bird into view in order to observe them more closely. These were all generalized reasons for feeding we had suspected and would be expected of feeders anywhere in the world. It was the second-most-frequently mentioned group of responses, however, articulated by almost 40% of feeders, that was entirely unexpected. Peter deliberated over what to call these responses before settling on the label "Atonement."[3] Based on the detailed written responses but especially the passionate expressions some people used to describe their concerns to us during the postsurvey debriefing, this was by far the most significant reason for their feeding. Typical statements were: "Humans have destroyed so much of the natural world, I am trying to give something back"; and "We have taken away their homes and poisoned their environment yet they still sing for us. It makes me ashamed of what we are doing, but putting out the food may help." Whether these are responsible or even sensible statements is beside the point. They are strong and powerful sentiments and evidence of deeply held convictions. *Sure, it's nice to see the birds up close, admire their colors and feel happy* (these people seem to be saying), *but there are also serious environmental issues to confront. I might not be able to stop human development but I can put up a feeder and the birds will come. At least I can try to do something!*

This modest but important exploratory study marked a significant turning point in my interest in bird feeding. I suddenly realized that this "mere pastime" was often much more than simply a way for people to "pass the time." Much earlier, my perception had been that bird feeding was a fairly casual, private, and passive affair, the equivalent of daytime television but without the contrived drama. To discover that large numbers of people were feeding and that many of them had strong, often ethical convictions and felt they were engaged in a form of environmental activism was a major revelation. Moreover, far from being passive, feeding for many participants involved direct and active participation. Above all, these people (admittedly a smallish sample from one small part of a country that doesn't even like feeding) seem to really care. Peter's study only hinted at the complexity that was to come.

The Many Dimensions of Motivation

It has taken a while, but there have finally been several significant investigations into the complex motivations of bird feeders. Unfortunately, despite the colossal scale of the bird feeding industry in North America and continental Europe, there has been relatively little research conducted on most aspects of what is probably the most important human-wildlife interaction in these regions. Certainly there have been excellent studies of who is feeding, where and what, on the movement and behavior of birds at and between feeders, even fascinating new work on the monetary value people place on birds in gardens.[4] To my knowledge, however, no one has yet investigated the motivations of feeders from most parts of the world. This means that we currently cannot be definitive about the particular motivations of feeders from anywhere other than the United Kingdom and Australia right now. While I am certain that many of the findings we discuss here will be of relevance everywhere, regional differences are certain to exist. Do the feeders of Manitoba feed for the same reasons as people in New Mexico, Lithuania, Malta, or South Africa? We await these future studies impatiently.

One important exception to the dearth of US studies of these matters is a remarkable investigation of people who specifically feed crows.[5] This is not typical garden bird feeding but is certainly worth mentioning in the current context. During his many years of studying American Crows around Seattle, John Marzluff encountered numerous individuals who regularly fed crows, often in particular locations and in very particular ways.[6] It slowly became clear to John than this was not a regular human-wildlife interaction; something special was happening. Teaming up with social scientist Marc Miller, the two carefully observed the often complex interplay between person and crow. The researchers distinguished two categories of crow feeders. One group, crow observers, were mainly interested in the behavior of the birds and were somewhat detached in their interactions. For the other group, in contrast, the researchers did not hesitate in naming as "crow friends." These people intentionally attempted to communicate with the birds through a wide range of gestures and vocalizations. And this was most definitely a two-way exchange: the crows also appeared to be attempting to communicate with the people. For these crow friends, offering food is far more than supplying a little sustenance; it also allows for shared gestures, intimate yet indirect connection, mutual regard. The authors make a provocative claim: that this is a form of genuine relationship, founded on but extending far beyond just feeding.[7] Relatively few of the species we feed seem interested in this type of commitment. I am not at all surprised to learn that crows just might be.

Our current understanding of the motivations for feeding garden birds comes mainly from just four investigations conducted within the past few years. Each differed in approach and offer quite varied insights. They also covered much more than motivation, though that will be our focus here. The first and simplest study, by Lela Schreiber, was based on an online survey of several hundred members of the British Trust for Ornithology.[8] This study was directly influenced by Peter Howard's pioneering work, with respondents being asked to nominate which of a list of six major reasons best explained their motivation for feeding. This forced participants to choose among some of Peter's categories as well as several others devised to be more appropriate to Britain. As with the earlier study, "Pleasure" topped the list, with 83% of respondents ticking this box. Significantly, the second-highest response (77%) was for "Aid survival," a topic not even mentioned in the Australian study. The third-most-frequent (60%) reason nominated was "Studying behavior." Peter's crucial "Atonement" category, reworded as "Making up for environmental degradation," came in fourth but was still an important concern for over 40% of respondents. Like the earlier study, this British work showed clearly that feeders were motivated by a broad mixture of influences, though there was also broad correspondence between the two countries.

Dave Clark was influenced by the two earlier studies, but he took his work into entirely new territory.[9] Dave is an experienced social science researcher with a background in marketing analysis. (He is also a keen cyclist, feeder, and birder who has guided me through the unexpected biodiversity of inner London on several occasions, as mentioned in Chapter 7.) When he decided to expand his skills and interests by undertaking a master's in ornithology at Birmingham University a few years ago, his lecturer and research supervisor, Jim Reynolds, suggested that Dave devote his considerable energy into discerning the mysteries of bird feeding as the research component of the course. This was exactly the sort of methodological challenge Dave thrived on. His plan was to incorporate both qualitative (subjective themes and meanings) and quantitative (all numbers and statistics) approaches in an attempt to uncover underlying patterns in the reasons people gave for their feeding. This was achieved by lengthy interviews with 30 feeders followed by an online survey, to which over 550 people from throughout Britain responded. To gain the necessary information from the interviews, it was essential to the qualitative analysis used that the participants not be influenced by the views of others (including the researchers) and that they were able to tell their story openly. From their responses, Dave was able to identify 8 different categories as reasons for feeding. These were then incorporated into the survey questions for the quantitative phase. Dave told me that while he was obviously aware of Peter Howard's categories, he was able to put these aside and allow the "data to speak for itself." "It's quite straightforward once you have some experience," Dave explained. "And besides, this was not much like Brisbane [the location of Peter's study]." Through the misty window we could see a lone bedraggled Blue Tit pecking at Dave's almost empty feeder, a cold drizzle settling in. I had to agree.

Dave's categories—the motivations or reasons given for feeding by bird feeders in the UK—are listed below in three groups indicating their relative importance.

Group 1: Pleasure; Bird survival
Group 2: Nurture; Children's education; Connecting with nature
Group 3: Making up for environmental damage; Companionship

These groupings offer a fascinating insight into the otherwise private attitudes and influences that lead people to feed birds. The two motivations that make up the first group—pleasure and assisting birds to survive—were entirely distinct from one another but were both of overwhelming importance to almost all the respondents. Or maybe it simply feels good to know you are helping birds survive. I suspect these elements would be very difficult to separate.

The motivations included in the second group were almost as significant to a large proportion of feeders but suggest several new and perhaps unexpected reasons to feed. For example, one respondent, identified only as "female, older" (defined as over 55 years) stated: "I live on my own, the children have grown up, I have no grandchildren, I have no job now that I am retired, I need something to nurture, to sustain life and watch life grow." Given the reality that by far the largest demographic engaged in bird feeding throughout the world is "older females,"[10] an element of nurture is undoubtedly going to be a major motivation. The role of feeding in the education of children had not previously been identified by other research but is clearly of great importance, and will be discussed in some detail later. Similarly, this was the first time that a form of "connection with nature" had been associated explicitly with the feeding of birds. This feeling or perception that people, especially those living in large cities, have somehow lost this connection with the natural world is of enormous concern at present. There is particular concern that this lack of direct engagement with nature among children may have serious implications for their future well-being as well as whether such children will appreciate the importance of conservation or value biodiversity. The feeding of birds may have a profound part to play in addressing this dilemma (see more later).

Interestingly, the so-called Atonement motivation appeared to have been a relatively minor reason for British people to feed birds, although around one-quarter of respondents still rated it at the highest level of importance. For whatever reason, the respondents in this study appeared to see their feeding as benefiting birds directly, through enhanced likelihood of survival, rather than through the less tangible "making up for environmental damage" idea. The final motivation, "Companionship," while of fairly modest significance, is both understandable and prosaic: feeding birds offers an important form of companionship, especially to people living alone. "To tell you the truth, I don't know what I would do without them. I have to keep feeding them so that they will visit."

How Do We Value These Visits?

The third exploration of the motivations we will discuss was also the most ambitious. Renee Chapman has been working with me on various bird-feeding projects for years, cutting her teeth on some of the first studies ever conducted on duck feeding.[11] But these were strictly ecological and observational: How many slices of bread? Which species got the food? What is the water quality in the pond like? Counts, percentages, graphs; all neat and straightforward. "I had been trained as an ecologist and knew how to record what I saw," Renee explains. But the more she watched the interactions between people and the birds they were feeding, the more she wondered about what these people were thinking and why they were doing it. "I eventually realized that I was actually paying more attention to the humans than the ducks! This was a bit disturbing because, after all, I was a red-blooded ecologist; I knew that people don't matter!" This was obviously hyperbole but marked a turning point in Renee's perceptions of what was going on with feeding. "This was not really about the amount of food or the species of bird involved. This was clearly about why people fed birds in the first place," she continued.

The pivotal moment in Renee's transition from ecologist to—well, something else—was the arrival of a tradesman at a pond at lunchtime (also described earlier in Chapter 1). I still recall her vivid colloquial description of this encounter, soon after the event: "Here was a tradie in his flouro gear [Translation: tradesman wearing high-visibility clothing], your typical tough-looking customer. He sat down on the grass and took out his lunch things. As he started to eat a sandwich, a couple of ducks arrived. I actually thought to myself: 'Good luck with this guy,' yet he then opened a container and started to hand-feed something to the closest ducks. I was astonished. When he had finished and was about to leave, I worked up the courage to go over and ask about the food he had offered. He said, proudly I think, "Cooked brown rice. Made it meself. Just for them. Lot better than bread. Me ducks seem to love it. Aren't they just great?'"

This was one of those little unexpected moments when things change, dramatically. Renee's tradesman interaction, which challenged some of her preconceptions and personal theories, sent her off in an entirely new direction in terms of her research. Now she needed to really get into the heads of feeders, to try and discern what drove people to feed birds. Thankfully we had on hand a couple of experts—Helen Perkins and Monica Seini—in the approaches appropriate for this field. Renee was soon on her way to becoming a red-blooded *ecological* social scientist.

Although it was clear that Renee would be investigating the motivations of feeders in Australia in greater depth than had been previously attempted, she was also keen to do something bigger. To me, there was a much more ambitious project waiting to be tackled: a comparison of feeding motivations between Australia and somewhere else. Not just to see whether there are differences—there will always be distinctions between any two places (a comparison technically regarded as "trivial")—but because of the strong contrast in community and societal attitudes to feeding. As described already, Australia is one of the only countries in which feeding is frowned upon and everyone is aware of it. Nonetheless, people still feed in large numbers. Does such a context influence the reasons that people feed, compared to a country where feeding is actively encouraged? And because of the earlier work already conducted there, the United Kingdom was the obvious "other" place. With the enormous assistance of Jim Reynolds and the British Trust for Ornithology, Renee was able to temporarily trade Brisbane for Birmingham, Blue-faced Honeyeaters for Blue Tits, parched paddocks for a green and pleasant land.

The data Renee used were obtained in ways familiar to us by now: a carefully developed online survey of hundreds of feeders from Australia and the UK, identical except for some culturally specific terminology.[12] However, as this information was to be interrogated for its content (using techniques known as content analysis), people needed to be able to express themselves freely; this was not a "Which statement best describes . . ." type of survey. The key question (Q6) for us (there were 31 in all) asked simply: "Why do you feed wildlife?" (Of course, "wildlife" implies birds, but there are also plenty of bird feeders who offer a little something to hedgehogs, possums, foxes, and lizards.) For many social scientists, such broad open-ended questions are usually avoided; people often don't know how to respond, we were warned, so many will simply will not write anything. It was a risky approach, but Renee didn't want to manipulate or influence the answers. Maybe respondents would complete most of the survey but skip this one. You need somewhere over half to be confident that you have an acceptable representative sample, we were told. When the survey closed, Renee had 647 usable surveys from Australian feeders and 212 from the UK, with 99% and 98% completing the key Question 6 respectively! This level of response is almost unheard of. It seemed that people really did want to tell us why they fed birds, so much so that their answers often went on and on. "I had the opposite problem to what I had initially been worried about: too much good stuff!" said Renee. It was a problem worth having.

With such a rich seam to be mined, it was time to delve in a different direction from that of previous studies. The various themes associated directly with feeding had been well described in studies from both Australia and the UK. Renee wanted to see how the detailed open-ended responses fitted with other dimensions of the relationship between people and wildlife. An obvious starting point for this kind of exploration were the "wildlife values" developed by Stephen Kellert from Yale University during the 1970s.[13] In what was truly pioneering work for the times, Kellert attempted to clarify the many different ways that the people in the United States interacted with, regarded, used, and cared about wildlife in all its forms. After an enormous number of interviews with people from all walks of life throughout the country, Kellert discerned nine different dimensions to this relationship. He called these "wildlife values" and explained rather technically that these were part of a multidimensional construct; any individual almost always "valued" wildlife in numerous ways and the relative significance of these could vary depending on the species and the situation. It was all too easy, he warned, to label someone as "being" of a certain value when this could change quickly. For example, consider a plausible relationship between a middle-aged US male and white-tailed deer, an abundant game species in the eastern United States. If this person was an experienced hunter, it would be easy to assume that he values deer in what Kellert called a "Dominionistic" manner, suggesting that his relationship with the species was primarily one expressing mastery of man over beast. In reality, our deer stalker might explain (if asked, though probably not in these words) that he, in fact, greatly values the deer from a "Naturalistic" perspective, having genuine respect for a magnificent animal in its natural world and gaining real satisfaction from the skills needed to successfully stalk his quarry. Once the deer has been dispatched, however, this hunter might value the venison thus obtained on a "Utilitarian" dimension of the relationship.

While Kellert's wildlife values have been very widely used—indeed both Lela Schreiber and Dave Clark employed them in their work—there has also been quite a lot of discussion and criticism over their applicability and misuse.[14] Much of this debate is over whether his values are transferable to non-Western cultures and other contexts. While these are valid concerns, they are of little relevance to our discussion here. In her work, Renee has explained how she adapted Kellert's values in her primary purpose of understanding the dimensions of the motivation for feeding in the two countries. Rather than describe all the findings in full, we will focus on the top four values found to explain the motivations in each country. Recall the very different societal backgrounds being compared. Renee and I are obviously disinterested and objective scientists, don't forget, but we were both wondering just how different feeders' motivations from the two countries were going to be. It's tits versus lorikeets, siskins versus kookaburras after all.

The first surprise was that the four dominant values (themes) describing the motivations of feeders were exactly the same in both Australia and the UK, although their order and significance were different. These values were Naturalistic, Moralistic, Humanistic, and Ecologistic-Scientific. Let's tease this apart. First, a clear majority (79%) of feeders in the UK were strongly motivated by "Moralistic" impulses. This motivation is characterized by a strong ethic of care and personal responsibility for the birds they feed. Overwhelmingly, British feeders really *care* about the welfare of their birds; they want to help. Within this broad category, Renee's analysis discerned two clear elements: motives of atonement and of support. Two quotations from UK feeders illustrate these sentiments. The first is that "the pressures placed on the natural world by humanity make it important for us to provide as much support to wildlife as possible." The second, that "birds suffer during the winter due to a lack of food. Feeding helps more of them survive the winter." The "Moralistic" motivation—helping—was also strongly evident among Australian feeders, though statistically lower in significance compared to the UK. Nonetheless, this value was second in importance in Australia, with 38% of feeders identified as such. This prominence was unexpected as most of the continent does not experience the harsh reality typical of Northern Hemisphere winters. Rather than annual winters, however, it was the less predictable but not infrequent climatic extremes—prolonged droughts, cyclones, excessive heat—so characteristic of Australia that seemed to motivate feeders. "I started [feeding] during the drought when there was no food or water," said one respondent. "Birds visited [towns] in flocks, scavenging. I have kept it up ever since." The most important motivation among Australian feeders, however, was classed under the "Naturalistic" value, and associated with 49% of respondents. Unlike the clear prominence (almost 80%) of the "Moralistic" value among UK feeders, there was a much more even spread of the main values in Australia, indicating that these were more equal in importance for most feeders.

The "Naturalistic" connotation is one of enjoyment and pleasure in the interaction, a feeling of personal happiness in the presence of these wild visitors. Not unexpectedly, this value was second in importance for UK feeders, being identified with the motivations of 44% of respondents. Interestingly, this was the only one of the four top values that was not statistically different between the two countries.

The other two values we are considering contrasted strongly in prominence between Australian and UK feeders. "Humanistic" motivations—which emphasize the relationship between feeder and visitor, the perceived connection with a wild creature and its importance—was much more significant among Australian feeders (27%) compared to those from the UK (11%). Again, this was something of a surprise, as was the depth of feeling evident among the responses. Consider these Australian examples: "To have the birds seek our company is salve for our souls," said one respondent. "There is a certain trust established and there is an incredible feeling to such a close encounter," said another.

The final motivation is the dauntingly entitled "Ecologist-Scientific," one of Kellert's original values that had been subject to considerable revision. For our purposes, this classification attempts to capture the more objective, dispassionate aspects of feeding, where the emphasis is on learning and observation rather than the personal experience associated with the "Naturalistic" or the overt caring of the "Moralistic" values. This orientation was much more prominent among UK feeders (38%) than Australian (15%) but, in both cases, was of much less significance to almost all feeders than the values associated with caring and relating. This aspect of the feeder's motivation adds an element of curiosity to the more emotional experiences of being a feeder.

There was much more as well—elements of aesthetic appreciation of the beauty of the birds and of the value of companionship, for example, but these were of considerably less importance to the majority of feeders in both countries. What this fascinating investigation tells us with great certainty is that many people are motivated to feed by a number of powerful internal influences, with the need to assist the survival of birds, along with the personal reward of the experience, being especially strong. In addition, feeders and their birds were often engaged in terms of both relationships and fascination. Australian and British feeders, despite the culturally differing backgrounds, were remarkably similar with regard to the four dominant dimensions of their motivations but differed markedly in terms of what mattered the most. For Australian feeders, all four were expressed without a clear leader, while British feeders were activated much more by motivations of care and support.

How Much Do You Care, Really?

Renee had another means of gaining even more insights into the personal world of the feeder. This approach was as overtly quantitative as the Kellert values were qualitative. Because of the predominance of caring, enjoyment, and engagement emerging from the earlier studies of feeders' motivations, Renee wondered whether there was a way to objectively assess whether feeding enhanced participants' perception of a connectedness to nature. The apparent loss of a sense of belonging to nature, or even the relevance of nature to contemporary human life, is becoming of increasing interest and concern. With the majority of people now living in rapidly expanding cities throughout the world, the opportunity to interact with natural landscapes or even to observe wild animals is becoming less and less possible. The world many of today's children will inhabit is likely to be even more ordered, filtered, and domesticated. Direct experiences of nature are progressively being replaced with indirect or even artificial replicas of the real thing. This process has been termed the "extinction of experience" and has been linked to all manner of future dystopias.[15] Why should tomorrow's citizens care about conservation or national parks when they have the David Attenborough documentaries on YouTube?

These concerns have generated numerous attempts to distill the essence of people's personal attitudes, beliefs, opinions, and actions into a measure—a specific number perhaps—that represents where they stand on a continuum of values. This is effectively the opposite of Kellert's multidimensional values approach but can be of great value in mapping the trajectories of societal views and perspectives on all sorts of issues. This is the realm of "scales," in which one's responses to a long list of statements can show you where you come in relation to others, or at least the 12 other people who have taken part in a particular study. (I recently found myself coming in at 0.52 on a 0–10 scale of visual art appreciation. I still don't know what to think about that.) In relation to what may be called our connection with nature, there are now a number of well-developed scales that purport to provide reliable measures of dimensions such as environmental awareness, concern for nature, and "belongingness." These scales include the Connectedness to Nature Scale, Inclusion of Nature in the Self Scale, and the New Ecological Paradigm Scale, each of which has been employed widely and, inevitably, scrutinized ruthlessly.[16] Some have proved to be accurate in predicting involvement in environmental activism and long-term commitments to demanding activities such as raising orphaned wildlife.

Renee had to wade her way through a vast and dense literature that explained, defined, critiqued, and sometimes ridiculed these scales, getting increasingly confused and frustrated. Eventually, she came upon a reference in an article to a very new scale that appeared to be heading in the right direction. Renee knew that she needed a scale that tapped into the intimate, undoubtedly emotional aspect that was salient in the "caring" and "relationships" dimensions of the feeding experience. This was crucial since most of the other scales appeared to be primarily aimed at cognitive dimensions, the more objective, factual, "science" elements. Although the name of this new scale, the "Love and Care for Nature Scale,"[17] was a little off-putting, it claimed to reliably measure "love and caring of nature, including a clear recognition of nature's intrinsic value as well as a personal sense of responsibility to protect it from harm." This was just what Renee was looking for, though we needed to be convinced that is would be suitable. At least as important, she needed to be able to use the scale herself, but some researchers were very protective, or demanded large sums to gain access to the test items that the participants respond to. Nonetheless, Renee decided to find out more.

The first shock came when we searched for the location of the author of this scale. Helen Perkins, we discovered, worked at our own university, though at a different campus. Instead of the prolonged and fraught negotiations we feared, a few days later Renee and I were chatting over coffee at Griffith University's Gold Coast campus, less than an hour's drive away. It was immediately apparent that this was going to be a mutually beneficial collaboration. Helen had developed her scale for use in ecotourism but was convinced that it had much wider validity across a broad spectrum of scenarios, anywhere that led to feelings of "awe and wonderment," as she explained. A brief outline of what Renee had already discovered about the bird-feeding experience was all Helen needed. "Let's do this," Helen declared dramatically. "Oh, and the name," she added. "Unfortunate, I know, but quite accurate. The original title was the 'Environmental Ethics Scale,' but that sounded too generic. Like it or not, the scale captures the strong emotions sometimes associated with really special experiences. It sounds like this is happening with at least some of your feeders. We will soon find out."

The arduous process of developing her scale involved Helen progressively testing the reactions of different groups of people to a series of statements. Participants indicated the strength of their agreement or otherwise by ticking one of 7 boxes between 1 (strongly disagree) and 7 (strongly agree). From an initial 100 statements, careful statistical evaluations of consistency eventually left a list of just 15 statements. These were then included in Renee's online survey described earlier. After all that preparation, the analysis was, after the usual checking for errors, fairly straightforward. And brutally sharp, the deep and passionate expression of care and concern feeders have for the birds they feed, reduced to a single number.

But a single number means nothing by itself. Its significance can only be assessed in comparison to other groups who have also responded to the same statements. Unfortunately, the fact that the Love and Care for Nature Scale was so recent meant that only a few studies have employed it; but these will have to do. First, three separate studies of different student cohorts undertaking university studies in business yielded the following scores:

Undergraduates	4.31
Master's candidates	4.94
International students	4.82

Given that the maximum score possible was 7, and that 3.5 would be utterly neutral, and that the respondents involved were relatively well educated but certainly not particularly pro-environment, we can probably regard these scores as fairly indicative of the upper socioeconomic end of the general community. As a contrast, Helen assessed a large group of people engaged in high-end ecotourism activities. These people had paid considerable amounts in order to travel to distant locations specifically to observe and interact with nature and wildlife. These folks would be expected to have a higher score on Helen's scale, and they did:

Ecotourists	5.36

Now we can validly compare our feeders. They seem to care; let's see how they scored:

Australian feeders	5.90
British feeders	5.80

This was, I have to say, very big news! While the results from all the preceding studies paint a picture of feeders being strongly motivated by a combination of compassion and pleasure, along with feelings of connection and a willingness to learn, the extraordinarily high scores from both countries on the Love and Care for Nature Scale point directly to something powerfully emotional.

Feeding the Connection

The final study appeared just as I was completing the very last sections of this book. I had been in contact with the researchers, Daniel Cox and Kevin Gaston from the University of Exeter in Cornwall, and had an inkling that they were working on something highly relevant. I was right. Their research adds substantially to our understanding of the significance of feeding to people engaged in it.

The context for their study was the important benefits that *people* can potentially obtain from some level of interaction or connection with nature (discussed further in the final chapter). While these have been associated with all sorts of fairly passive activities—viewing natural landscapes, walking in parkland, listening to natural sounds—Cox and Gaston proposed that maintaining a bird feeder may be something much more immediate and direct—and therefore possibly more effective in enhancing feelings of well-being.[18] They wanted to see whether the commitment associated with having a feeder led to feelings of well-being and whether a higher level of engagement (feeding more often) had correspondingly greater results. These questions were investigated among the general public—not just people who fed birds—in three English towns: Milton Keynes, Luton, and Bedford. And unexpectedly in this age of online surveys and research by virtual role play, perhaps in keeping with the theme of direct connections, the researchers reverted to the oldest social-science research approach in the book: knocking on doors and chatting with people face to face. Amazing.

In the end, they had data from 331 real but randomly selected people, of whom 83% fed birds (recall the figures presented in Chapter 1; even Cowie and Hinsley[19] would be impressed!). To cut to the chase, people who actively engaged with feeding reported feeling more relaxed and connected to nature as a direct result of this activity. Moreover, these perceptions increased with the level of feeding undertaken, and especially when the birds were actually observed at the feeder. Although they make the obvious point that they were not investigating how the apparent relationship between feeding and feelings might come about, the researchers argue that if people are willing to report this connection, something is clearly going on. This is a great starting point for a future physiological investigation. Meanwhile, the humble bird feeder has acquired a new status: "A bird feeder has the potential to be a powerful tool for people to make this connection [with nature] because it provides a focal location where people can expect to observe birds."[20]

This commonplace, homely, possibly even quaint pastime is, it seems, also a deeply felt *experience*. The caring element associated with feeding seems to be much more than just a transitory whim; you can't simply manufacture this depth of emotional attachment. That is not to say, however, that feeders are simply sentimental suckers, manipulated by clever birdseed marketers (as well as by the birds themselves). It would be rather naive to think that such influences are not a major element in the tremendous success of the bird-feeding industry. But I am starting to think that it is probably the other way round: the industry works so well precisely because people already care so much. Even the most sophisticated marketing campaigns fail if there is no emotional engagement. "We only care about what we love."

Previously I described an early perception of mine of feeding being something akin to the passive watching of television. In view of the complex, powerful, emotional, important, even profound insights gleaned from the research discussed in this chapter, it is clear that such attitudes—not at all uncommon—were based on my own extremely inadequate understanding. I was wrong. This is not just a way to pass the time: bird feeding really matters!

9

Bird Feeding Matters Even More Now

The Promise and Risks of a Global Phenomenon

The birds at my table are *very* impatient. I am just back from yet another trip and there has been no food on the feeder for an entire week. That was deliberate. It was a difficult decision to make, but in the end, I decided not to arrange for anything to be provided while I was away. Yes, I know (I can feel the e-mails and tweets building up already), I broke the feeder's Golden Rule: *Once you start, don't stop*. And I did so intentionally, perhaps even defiantly. Was this blatant cruelty, willful neglect, or straight-out stupidity? Don't I care about "my" birds after all? After I upend the cup of wild bird mix and retreat inside to watch the lorikeets jostle and grumble as I sip my coffee, I reflect on the unforeseen dangers of acquiring knowledge. Sometimes the things we learn can lead us to reconsider our ideas and maybe even our actions.

This has been a long and fascinating journey, metaphorically, emotionally, and physically, including a lot of air miles. Along the way I have been challenged, astonished, appalled, and uplifted by what I have seen and heard. As is so often the case, when I started I thought I didn't know much about the topic; now I *know* that to be true. Even though I have gathered together enough material to fill a sizeable book, I am even more acutely aware of how much we don't know about the feeding of wild birds. There may be hundreds of books describing how and what and where to feed in many countries, but the amount of genuine scientific research investigating what this means, for people, ecosystems, landscapes and the birds themselves, is shockingly inadequate. We have barely scratched the surface of the many vital questions that need to be answered. I think that this is changing, and I hope this book pushes things along. You see, I think that there is a lot at stake here.

What we do know may be from only a few locations around the world and may be fairly limited in scope, but many of these ideas and principles can probably be applied to gardens and feeders everywhere. Let me illustrate. Take my feeder in subtropical Australia, right now a scene of intense consumption of sunflower seeds and white millet. Wait, there appear to be some other items on the feeding platform that I did not provide: in my absence, it looks like quite a lot of tiny seed capsules from the eucalyptus trees above have collected on the feeding platform. I vaguely noticed them when I put out the seed just now, but they didn't trigger an acute "hygiene alert," so I took no further notice. But the lorikeets certainly have, and are now concentrating on these capsules, deftly splitting them with their wickedly sharp beaks to get at the minute seeds within. There is a distinct irony in this scene: these wild visitors, who came looking for birdseed, have ignored the commercial stuff (which I paid good money for) in favor of the natural bounty they have unexpectedly discovered. Natural food in an unnatural setting.

This rather unsettling observation event may help me explain my simple though significant decision to withhold provisioning while I was away. Among the vast amount of detail I have carefully considered in constructing the contents of this book, I have kept particular note of things that relate to my own practice and ideas as a feeder. I hope that you have too. For example, the exploration of the various reasons that people have for feeding got me thinking about my personal motivations. I would say that the wonder and pleasure of seeing truly wild birds up close is paramount for me, though as a professional ecologist interested in behavior, I am also fascinated in their interactions. But I am willing to admit, nonetheless, that an emotional element is also strong: I would score (*do* score, actually;

Renee tested me) pretty highly on the Love and Care Scale (just don't tell my wildlife colleagues). So, yes, I do care; I care greatly about the well-being of the individual lorikeets that visit my feeder (and I have also started to care deeply about their cousins elsewhere currently consuming far too much meat). But caring also means that I want the best environmental and ecological circumstances for these birds to live full and productive lives. For this species, in this landscape, the provisioning of a little seed is probably of very little consequence. I know that there is a huge amount of natural food available for lorikeets (and many other species) throughout the local landscape and that it is readily available for most of the year. By limiting the amount of food I provide, I hope I am encouraging these birds to continue to live and feed as naturally as possible.

That is now. Things can change suddenly, however. In recent years, our region has endured several spectacular storms that stripped the trees of their flowers and fruit at the peak of the breeding season. Everyone reported more birds than normal at their feeders and birdbaths over the following weeks. A decade ago, we experienced a sustained and brutal drought that led to the death of many plants and withered the foliage of even ancient trees, greatly reducing the availability of natural food and standing water over a huge area. Birdbaths, not that common in this normally wet region, became all the rage, with influxes of species from hard-hit areas turning up in places they had rarely been seen before. Caring also means being ready to respond when needed.

Is Caring Helping?

Some questions:

What if there isn't a lack of natural food?
Are birds really reliant on well-provisioned feeders?
Does feeding actually help birds, other than during difficult times?
What if feeding is assisting species we don't want or is hindering those in trouble?
In fact, should we actually be feeding at all?

That's being provocative, obviously, but consider your reaction to these questions for a moment. Are you shocked or disappointed that anyone could seriously suggest that feeding wilds could be anything but positive? Have I just been maliciously leading you along this path only to spring a trap and suddenly announce that feeding is fundamentally bad? If you have been following the story to this point I hope that you will expect that such a pronouncement is unlikely. It should be quite clear, however, that there are some serious issues to be faced and that every individual person engaged in feeding should be willing to consider where they stand.

I pose those particular questions because at present we simply don't know or there is no sensible answer. These are, however, the kinds of question that are almost never asked yet underlie much of the success of the bird food industry. We are surrounded by messages that proclaim that we need to feed birds because they need our help, or because we need to make up for lost natural resources, or even that the commercial products are better than what is available in the apparently depauperate landscape. Or the message may be simply that supplying birds with a little something is just a nice way to see these gorgeous creatures. We respond with our purchases because we care and this seems like the best way to help. And we continue to do so because it is so rewarding: the birds really do come. It's almost as though they come as a way of acknowledging our efforts and are showing their gratitude. It's a simple equation and appears to be clearly a win-win transaction, at least in terms of the suppliers of the products and the people buying the stuff. We care, they offer, we buy, they prosper, we feel great. The wheels of a gigantic industry roll on, profits continuing to grow annually.

I can be a little caustic about this, safe in the knowledge that the bird food industry is here to stay, for all the obvious reasons. This is perhaps the perfect business model, with consumers utterly convinced about the value of the products, compelled to buy as a way of doing good, and propelled by a powerful emotional engagement in the process. We buy because we care, leaving the companies free to dream up next season's must-have Organic Portuguese Raisin-Flavored Chickadee Supreme with added Vitamin H and a hint of MiracleMineral. Or something like that.

But have you noticed anything missing from all of this? Yes, the birds. The industry can do what it likes (and it does), but in the end what we, as feeders, must be entirely focused on are the birds and what all this supplementary food means for them. Let's ask a much simpler question:

Does my feeding help the birds that visit?

This is deceptively simple, of course. While we are all likely to respond immediately and adamantly yes, just how can we be so sure? How can we ever really know whether even our most sincere and committed (not to say expensive) efforts, motivated by pure care and concern, are actually doing any good? Apart from some extremely specific studies on survival rates reviewed here or elsewhere, and despite the scale of this practice, there is virtually no information that can confirm that our caring actions are actually helping in any real way.

Thankfully, there is plenty that we do know. There is a large body of research on many important and relevant issues, as summarized in this book. While there are many things we will just have to accept as unknowns, especially in the urban and rural landscapes where the bulk of feeding occurs, there is much that is useful and instructive.

What Do We Know about Feeding?

As we approach the conclusion to this exploration, I have attempted to extract some key points from the considerable amount of material we have discussed so far. These points are presented as what I regard to be factual statements that emerge from all the studies and discussions I have presented. Some are simple statements of reality, others are more provocative. My aim is to try and take us from our natural focus on our feeders out there in the garden to their role in a much broader landscape.

1. The amount of food we offer to wild birds, and the amount consumed by them, is astronomical.

The figures are all there (see Chapter 1), and frankly, they are hard to believe. The image I keep coming back to is the line of 22,000 railway carriages full of the seed supplied to birds in the United States *every year*![1] And although it is all eaten (though not always by the intended critters), every single seed is *supplementary* to the diet of the birds. In other words, additional, subsidized, possibly not necessary; most of the time the birds don't actually *need* any of it. In the case of North America—or Australia or South Africa or perhaps most other places where people feed birds—there is absolutely no compelling case that natural food supplies are seriously lacking. Feeding to replace lost resources is much more plausible in parts of the world where feeding has a genuine conservation connotation. Certainly, for the more developed areas of Western Europe (Denmark, the Netherlands, southern England, and Germany, for example) gardens play a vital role as habitat for many hard-pressed species, with feeders an important component. In such places, feeders in gardens may very well be keeping some species or populations alive.[2]

But what about winter? Obviously food is seriously limited when it is cold, and the birds that don't migrate away clearly find it hard to survive. Providing relief for the suffering seems to be where all this began, and it remains the most pronounced motive for feeding in the Northern Hemisphere. Yes, these birds do need food in winter and people tend to respond. This is an entirely different scenario, however, involving assisting individuals for welfare reasons rather than whole populations for conservation. Survival through the winter can be significantly enhanced by having access to supplementary food, so more of these birds are likely to make it through to the following spring. These survivors will include some that are sick, maimed, infectious, and impaired in some way. Normally, the natural process of winter's demands would have removed these individuals from the population, leaving only the strong and tough to survive. With access to food, however, birds that would normally have expired may get to breed. Feeding may, therefore, be canceling out what was a "natural" (in the sense of natural selection) process, with potential evolutionary implications for the species. See what a simple feeder can do!

2. The bird food industry is relatively new and extremely influential.

People have always fed wildlife spontaneously, and even systematic, planned, organized feeding has been around for quite a while. Traditions of suet balls, feeding tables, and peanut hangers, for example, go back centuries. Virtually all the food and apparatus used was homemade or derived from existing supplies: food scraps from the kitchen, grain from farm supplies placed on simple feeding platforms whipped up for nothing in the home workshop. The arrival of mass-produced, ready-made, packaged bird-food mixes and feeders utterly transformed a private domestic activity with negligible monetary cost into a gigantic industry based on unbelievable amounts of money being paid for seed and associated hardware.

This has happened remarkably quickly, with much of the growth occurring since the 1980s.[3]

The influence of the big corporations involved in the bird food industry is not simply strong and pervasive. In many ways, these commercial entities run the show from top to bottom. From the beginning, their entirely legitimate financial objective of selling products and enhancing demand has driven every aspect of the bird-feeding movement. Industry marketing has always been extremely effective in recognizing the motivations and meeting the needs of the feeding public—well ahead of the researchers—and in the process has built an extraordinarily successful marketplace where consumers and producers interact with mutual satisfaction. Provided that nutritional standards are preserved, quality is rigidly maintained, and their products are rigorously assessed (that they don't add pesticide would be nice), it is hard to fault it as a successful model for business. Those are major caveats, however, and there are some very big and conspicuous exceptions. Nonetheless, the suppliers appear to be meeting (as well as manipulating) the demand of eager customers in innovative and imaginative ways. And let's face it, visiting a bird-feeding outlet can be a real pleasure, full of motivated people sharing stories of their latest visitation. Perhaps the industry's greatest triumph has been in convincing the buying public that they are actually interested in birds rather just selling things.

3. The move from feeding only in winter to year round is profoundly important.

Winters are tough, and small nonmigratory birds are unlikely to make it through unless they have access to a reliable food supply. Away from the city, this might be fat from a dead animal, insect pupa, the last fruit and seeds, or the suet balls in a farmer's garden. These days, the suburbs are full of feeders, topped up regularly right through winter by caring people. In many places, however, these supplies end when the warm weather returns. Indeed, feeding birds was traditionally a winter-only activity, humane assistance to struggling birds in times of trouble. Feeding in summer was thoroughly discouraged.

This traditional culture of feeding seasons is changing, however, dramatically and rapidly, especially in Europe. In many countries, feeding now occurs year round and is actively promoted, not only by the people selling the seed but also by the bird and conservation groups. Of course, feeding also occurs in places without serious winters, generally in the Southern Hemisphere where the reasons for feeding have always been different. In general, the size of the industry in such countries is far smaller. Throughout the Northern Hemisphere, increasingly, the feeders are full all year round. This is both an unparalleled cultural shift as well as a monumental ecological alteration. It's too soon to know what is going to happen next, but I think we can expect fundamental changes to entire bird communities and possibly even ecosystems.

4. Even a little supplementary food leads to change.

While research into the implications of bird feeding in suburban environments is still fairly limited, scientists do know an awful lot about the effects of what adding food can mean for birds in natural areas. Hundreds of careful experiments have shown conclusively that even small amounts of additional food usually lead to some sort of change in the lives of birds.[4] Although many alterations in activities and behavior have been noted in a wide selection of species, typical changes include starting breeding earlier, breeding more often, and producing more young. When you include an increased likelihood of surviving through to the next spring, one rather obvious conclusion drawn from these short-term, small-scale, strictly controlled experiments is, quite simply, that more food means more birds. Sometimes a lot more birds.

More birds might sound like an excellent outcome, especially if a species is struggling. This is just what the numerous conservation programs that employ supplementary feeding hope to see, after all. In the suburbs, however, the recipients are almost always regular garden birds, and these are not being supplied with, say, 30 grams of sunflowers every second day for the two weeks leading up to nest construction, or whatever the protocol being followed in a carefully monitored experiment. No, indeed, these birds are likely to have unlimited access to as many feeders as they like, every day of their lives! Let me say it again: the wild bird–feeding phenomenon we are engaged in is a gigantic supplementary feeding experiment on a continental scale.[5] But without carefully devised protocols or careful monitoring. And no end point. And because this experiment started decades ago, we have no way of knowing what the outcome has been or will be.

5. Feeding changes entire ecosystems.

Big call? I don't think so, not any more. It doesn't take much to link the logic of the preceding statements to the suspicion that all that food must be having a fundamental influence at the ecosystem level. Several important studies on the relationship between the amount of supplementary food and the abundance of local birds have demonstrated clearly that there are many more birds in places with lots of feeders.[6] The density of birds is closely correlated with the density of feeders. Interestingly, because the number of feeders is often related to the level of affluence of human communities, poorer areas are also likely to have a lot less birds; there is less cash to spend on feeding. This is important because less well-off neighborhoods already tend to have less access to the benefits associated with contact with nature (see below), such as green space. Mind you, in the UK at least, the lowest levels of feeding have been found to be in the richest areas.[7] But while not really unexpected—birds are more likely to hang around in areas with plenty of easily accessed food—this relationship between feeders and bird abundance does again confirm the profound, landscape-level influence that is possible due to a simple, private action. It's just another feeder, but it may be one more rather commonplace illustration of the cumulative impact that the individual decisions of many people can have over vast areas. (Google "Anthropocene.")[8]

Birds have always moved around in response to the availability of resources, of course. It's the same with feeders. Although plenty of people fill their feeders daily, most do not, and many are irregular suppliers at best. With their survival contingent on the waxing and waning of seed crops, berry production, and beech masting, and on the occasional fortuitous bonanza (an overturned grain truck or a dead moose, for instance), birds everywhere have survived by being opportunistic. They take advantage of what is on offer, then move on when it's gone. Or maybe not if the supply is predictable and reliable, like a feeder that is topped up daily. But even regular sources can suddenly dry up or shut down. People with feeders eventually stop, leave, retire, die. When that happens, the birds usually just move on. If they don't, they would starve, but the well-publicized concern that birds may become dependent on our supplies have largely proved to be baseless. Apart from times when conditions are obviously challenging—prolonged snow, excessive heat, drought—the birds almost certainly don't rely on our feeders. Several decisive experiments, involving shutting down the food supply after a long period of use (in the most famous case, after 26 years), found no evidence of impact on the birds that had been using the feeders, even in winter.[9] The birds simply moved on. (Remind me again about that golden rule . . .)

It has taken a while to convince some people, but cities really are a rather special form of ecosystem. Admittedly, there are some dramatic differences between urban ecosystems and the natural sort, but they can be studied in much the same way. Apart from the obvious feature of being utterly dominated by the presence and activities of one particular species, urban ecologists are starting to reveal that, far from being nature-free zones, many cities actually support remarkably high levels of biodiversity. In part, this is because the original settlements were usually sited on fairly productive land, places were food could be grown to supply the growing community. And although the process of urbanization inevitably wipes out much of the local biodiversity, once gardens are planted, parklands established, and reserves of the original landscape declared, animals of all kinds move in, from outside the township or from the reserves within. They will find a very different type of environment, however: noisier, smellier, and full of people and their supposedly tame companion predators. Only species with certain characteristics are going to prosper—tolerance, cleverness, cunning, and aggression are all advantageous qualities—but urban environments also offer some definite advantages for those game to try.[10] Food is fundamental, and because people tend to like gardens, the associated fruits, berries, seeds, and, of course, insects, are all there for the taking. It is not always appreciated, but well-established suburbs typically have far higher numbers and species of plants than the surrounding forests or woodlands. Add to these "natural" resources the huge amount of food provided by feeders, and it is hardly a surprise that urban environments support plenty of birds.

Plenty of birds but not necessarily plenty of species. Studies that found that the density of feeders predicted the abundance of birds also found that there was no similar relationship with the number of different species. In other words, all that food had little effect on diversity. In natural ecosystems, complex food webs can be constructed showing all the ways that myriad species interact (read "consume") with their food types: granivores, fructivores, carnivores, scavengers, and so on. The most stable ecosystems are also the most complex, and the most resilient to disturbance. We have always known that urban ecosystems are quite different, obviously, but recent detailed investigations of some urban bird communities found them to be extremely "unbalanced" and therefore potentially susceptible to sudden change.[11] Such systems, theoretically, have low resilience. The reason: an unusually large proportion of the species are associated with a single food source: bird food.

6. Feeders attract feeder birds.

You may be lucky and have a lot of different species visiting your feeders. I recall walking around a large snowy garden in Ithaca, New York, and counting more than 20 different species on the numerous feeders. When I asked why there were so many species in one place, my host, Natalie, explained. "If you want diversity, provide diversity. Here we have nyger, black sunflowers, sunflower hearts, safflower, white millet, suet . . ." Each feeder was attracting a different clientele and there were enough feeders for the visiting birds to be able to avoid one another. I think it was the greatest diversity of feeder birds I have seen, though I have been told of gardens where you can see even more.

These numbers are notable because they are extreme exceptions. Generally, a typical feeder attracts relatively few species, depending on its location and the current climatic conditions. It will be news to no one but does need to be said: our feeders are utilized only by species that are attracted to the type of food being offered. In most parts of the world, this will be seeds of various kinds, although just a few—black sunflowers being the leader—have dominated the market for quite a while. Seeds, as singles but even more so as mixes, are sold in vast amounts simply because many local garden bird species are granivorous. There are plenty of other sorts of food on offer as well, including suet, sugar mixes, and mealworms, for example, although these may been highly seasonal (such as suet) or species specific (the sweet solutions loved by hummingbirds, for example). In terms of the overall food supply across the landscape, however, it really is dominated by seeds.

The visitors that attend our feeders are certainly not a representative sample of the much more diverse community of local birds. That is not necessarily a problem; we all realize that there are species just too shy or specialized to be attracted to a feeder. Insectivorous species rarely visit feeders, even when mealworms are on offer (in many cases, the seed eaters tend to eat them first anyway). It does mean, however, that all the different feeders in the world cannot pretend to benefit the whole local bird community, as is sometimes claimed. And this can be a problem if some of the feeder species actually drive others away. Certain birds are naturally belligerent and will often roughly displace other, less pushy species. Usually, these victims tend to wait nearby for the bullies to leave before resuming their feeding (assuming anything is left). Sometimes, however, some species, typically those that are highly social, may occupy the vicinity of the feeder or an entire garden for long periods of time, effectively keeping all other species away. And sometimes the species attracted are not wanted at all. In a recent New Zealand study, for example, feeders were found to significantly increase the local abundance of introduced species such as House Sparrows and Spotted Doves, while some native species that didn't even use feeders declined.[12]

7. Most feeder birds eat mostly nonfeeder food.

Sometimes we seem to be convinced that our well-supplied feeders are almost solely responsible for keeping a lot of the local birds alive. After all, think of the time, money, energy and care we have invested in seed, feeders, and the like. I am sorry to have to break it to you, but this is extremely unlikely. Indeed, you may not want to hear it, but most of "your" birds also visit twelve other feeders around your area. Perhaps even more disturbing news is that, apart from some critical periods of physiological stress (such as when females produce eggs and during severe cold), or when certain key natural foods are extremely limited (late summer can be dire), the diet of most birds most of the time is not obtained from feeders at all. Sensibly, almost all birds, including the apparently more specialized, seem to have a rather diverse diet where possible, with a sizeable proportion of insects typically being included. The very few studies of the overall diet of several small bird species indicate that a wide range of different food types is necessary in order to cover all their nutritional bases.[13] These requirements change continuously through the seasons and with the various stages of reproduction. Sometimes they need more protein, sometimes calcium, sometimes some obscure micromineral found only in thistle heads. Even a good supply of Premium WildBird PowerMix will never provide all the nutrition that is needed.

This is important news with a humbling conclusion: our birds almost certainly don't need us as much as we need them.

8. Feeders do spread disease.

I have learned an enormous amount during this project but nothing as disturbing as the spread of horrifying bird diseases described in Chapter 6. There is no shying away from this undeniable fact: feeders have been directly involved in the rapid dissemination of House Finch conjunctivitis in the United States, trichomoniasis in the UK, and almost certainly other major epidemics among feeder birds. Feeders represent a textbook example of a perfect model of rapid and sustained contagion: highly infectious individuals are able to mix closely with concentrations of potential recipients with continuous mobility among many other similar locations. Once this process gets going it is virtually impossible to stop. With no effective treatment, House Finch disease simply burned its way through the population, resulting in the deaths of millions of birds. Trichomoniasis in Great Britain reduced the national population of Greenfinch by around a third over just a few years.[14] Wildlife disease experts frequently note that it is astonishing that there have not been many more similar epidemics. Future catastrophes are almost inevitable.

Unless we all take this component of our role as feeders extremely seriously. Everyone knows that they should be diligent about cleaning their feeders. Besides, we never see any sick birds at our place. And that is the normal, daily experience. Even during serious outbreaks, most sick or dying birds disappear fast; the lingering House Finches or the conspicuously pox-encrusted Greenfinches are notable exceptions. One of the paramount impressions I hope that feeders will take away from this book is that their garden—or balcony or window ledge—and its feeder are part of a vast, interconnected network of dozens or hundreds, maybe more, across the landscape, with birds moving endlessly between these points. Feeders are points of bird-to-bird contact, as well as sources of food, and therefore they are ideal places for picking up infection. We can't see this web of exchange or imagine the complexity of the bird traffic involved, but we do need to stop and appreciate both the extraordinary invisible network of contact all our feeders have brought to the world. Such a network simply did not exist before we invented it. The World Wide *Wings* Web, almost. And, a lot like the Internet, interchange is now possible on a truly grand scale, and with that, so are genuine risks.

I have been somewhat dismayed, therefore, to see how often this risk of disease spread is downplayed. With each outbreak, there is typically a campaign to reassure feeders that there is no need to worry. Certainly, keep an eye out for sick birds, and don't forget to clean those feeders now and then, but whatever you do, don't stop feeding! It's hard to determine what motivates these reactions. Just general reassurance to a concerned public? Another sign of a decidedly defensive feeding movement? Industry concerns that sales may suffer? I really don't know, but anything that attempts to minimize the responsibility of feeders to maintain the highest standards of hygiene is entirely irresponsible. Call me a bleeding heart, but it must always be about what is best for the birds.

9. Feeders care to see for pleasure.

Finally, real hope. I am still pondering the many implications of the research discussed in the previous chapter. Although people clearly feed for all sorts of reasons, to discover that the predominant motivations are care and pleasure and that these translate into a powerful emotional connection is enormously reassuring. It also explains in large part why so many people are engaged in this activity, are so committed, and why people from a wide range of countries around the world are involved. People are feeding because they want to help birds that they care about, whose visits bring them real pleasure. Simple. No, actually, it's rather complex because the mix of influences and feelings and impulses will be different for every individual. But caring and enjoyment appear to be particularly salient.

Why reassuring? Because this combination of strong emotional connection to something people really care about is an excellent platform for change. People who care deeply about their birds are much more likely to do what is necessary to minimize the spread of disease, for example. Or to think about the quality of the food they offer. Or to appreciate their part in the much bigger picture that is global bird feeding. People who care will want to learn and then respond. It's why this simple activity might actually be revolutionary.

Feeding Birds Can Change the World

A few years ago I organized a symposium on wild bird feeding as part of a big international bird conference. Among the people invited to participate was Rich Fuller, an English scientist now working at the University of Queensland, just across town from me. I was particularly pleased that Rich could attend as he has been involved in a number of the most important and influential studies into bird feeding ever conducted, delving into its patterns, practices, and implications in the United Kingdom. His research with a group of colleagues, mainly from the University of Sheffield and Exeter University, was part of a much broader investigation into the interactions of people and nature in urban environments, with a strong emphasis on understanding how ordinary people might be able to have real impacts on biodiversity conservation by the simple things they can do in their own gardens. The potential role of feeding wild birds was one of the key activities that might make a difference, given that private gardens make up one of the most important and extensive bird habitats in Britain. The studies were impressive and illuminating, of a depth and diversity unparalleled anywhere else in the world. Needless to say, Rich's ideas and findings have been cited everywhere in this book.

The speakers at our symposium presented talks on a wide variety of contemporary studies to a large audience to whom the scientific study of bird feeding was fairly new. These included Josie Galbraith's pioneering studies of feeding in New Zealand and the first outing of Renee Chapman's comparison between Australia and the United Kingdom. Most of the people at the symposium seemed fairly surprised that there was this much scientific interest in what they probably considered a pretty mundane topic. As each of the speakers progressively constructed an ever-expanding picture of the enormous scale of the activity, its geographical reach, and some of the major ecological and societal consequences involved, the change in the audience's attention was marked. *Gee, I had no idea this stuff was so interesting or important*, seemed to be the impression.

Rich was the final speaker, and the room was ready for some serious heavy-duty science. The hallmark of Rich's work is his powerful, sometimes dauntingly sophisticated analysis. After all, he had modeled a wide number of the socioeconomic, demographic, and geographical factors influencing participation in bird feeding for the whole of Great Britain, then synthesized this into almost readable scientific articles. But rather than hit us with facts, graphs, and stats, Rich simply walked to the center of the stage, asked for the lights to be turned up, and told us a story from his childhood without props or visual aids. He related how growing up in the suburbs of inner London had not exactly been the best preparation for a keen young birder, but he got to know the best places in his local patch fairly well. And while happy enough spotting the modest number of usual species, he dreamed of seeing some of the more exotic species that he knew existed farther afield. "My universe was fairly constrained, and of course I yearned for a taste of the wider world of nature. But our modest back yard in the middle of London was hardly likely to yield anything that surprising," Rich explained. He doesn't remember how it came about because his parents had no interest in birds or feeding, but somehow he was able to set up a simple feeding platform with seed borrowed from his caged budgie and even a mesh bag of peanuts to hang from a branch. It wasn't much but it did bring the tits and other locals into his garden where he could watch them up close.

And then came the day that changed everything, when suddenly anything seemed possible. "It was a cold winter's day and I went out early to check the feeder. A couple of small birds were pecking at the peanut bag hanging nearby, and I knew immediately that they were not species I had seen before, even in the woodlands nearby. I watched them closely for quite a while, almost disbelieving my own eyes. They were Siskins, a species I knew were normally found in the pine forests of Scotland. To see them there in London was just extraordinary. How could something so unusual, to me, so exotic, have discovered my little bag of peanuts? At the time they were very unusual for the south of England, let alone at my place in the middle of the suburbs. To this day I recall that feeling of wonderment and delight. This experience really did change my life in certain ways. This was probably the origin of my lifelong interest in thinking about biodiversity in urban areas. And most certainly, it made me realize, even back then, that such a simple act—putting out some food to attract birds into your garden—can actually make a difference to the lives of *people*. It made me realize that life and nature is not that far away after all, even in big cities, and that can be a life-changing gift."

While the symposium was an overwhelming success, and much of the information presented had been striking and informative, it was Rich's data-free, heartfelt revelations that everyone remembered and that have continued to be talked about ever since. To decide to eschew the conventions expected of a scientific presentation was always a bit of a risk, but Rich told me afterward that he discarded the PowerPoint in favor of the personal because that was the essence of his main message: it *is* personal. Even a little emotional, that scary concept so disparaged among the supposedly objective and professionally disinterested scientists among us.

Rich has since moved further along this path, considering the various ways that people interact with nature and the importance of encouraging this, particularly among the increasingly large proportion of the global population now living in cities. As we move further away, both physically and psychologically, from natural environments in the places we live and work, the opportunities for engagement with nature appear to be diminishing. We are spending much greater amounts of our time inside. Children are much more likely to recreate by engaging with digital landscapes and less time in the open air than even a few decades ago. Even when children are outside, this tends to be during times of organized sporting or recreational routines, typically with close adult supervision. City kids tend to have extremely limited opportunities for unstructured time outside to roam and play. There may be genuine parental concerns over perceived risks (involving strangers, accidents, animals, etc.), but an important consequence of this tendency, something that seems to be occurring worldwide, is a lessening of direct contact with nature. As mentioned earlier, this has been called the "extinction of experience," inferring that perceptions of nature are more often derived indirectly rather than through tangible personal interactions or experiences, through television, the Internet, Grandma's childhood stories, or even not at all.

We could go on to describe the many usually dire predictions about generations without this type of experience, as has been canvassed brilliantly by Richard Louv in his influential book, *Last Child In the Woods.*[15] Instead, let's consider some of the remarkable discoveries made about the benefits of contact with nature. This relatively new area of multidisciplinary research is verifying what many of us have always known intuitively.[16] For example, it has long been appreciated that the presence of companion animals appeared to calm and relax people, but the health and psychological benefits are now being measured and monitored with remarkably positive results. Many retirement villages, hospices, even prisons now have regular sessions where patients and inmates can simply experience the companionship of a gentle dog. Even such commonplace features as the presence of houseplants or fish tanks, views of natural landscapes, pictures of wildlife, and recordings of birdsong have been shown to reduce stress, speed up healing, and lower the number of medical interventions needed by patients in hospitals. Probably the most well studied of these concepts addresses the benefits associated with having access to green space.[17] For people who live in apartment towers or work in the skyscrapers of large cities, an increasingly large number worldwide, simply being able to walk to a nearby park or reserve has been shown to greatly reduce anxiety, workplace conflict, and the prevalence of many medical conditions and lead to increases in a range of measurable states of well-being. Perhaps the most astonishing result to date, however, is recent neurological research into changes in brain structure—specifically, the connections among neurons (neural pathways)—for subjects who walked through a forested park compared to those who walked along a busy road. Not only did the relatively short walk in the park significantly lower stress and improve concentration, these effects were correlated with an increase in nerve connections within the brain. If you live in the city, it seems, just being "in" nature can change you mentally as well as physically.[18]

That's the rather passive route to connecting with nature. With measurable effects even from such seemingly limited experiences as listening to natural sounds or wandering through a park, the next dimension of research into this realm is investigating how city dwellers can engage more deeply with natural environments and what the associated benefits are. Much of this work is focused on children, and there are now a plethora of programs in many countries currently under way that determine the numerous benefits of activities such as camping, bird watching, canoeing, wood carving, damming streams, and making mud pies. A worldwide movement, largely inspired by Richard Louv's revelations, is showing that tactile interactions, free-form, unplanned and spontaneous opportunities for creative expression—something technically known as "play"—has enormous benefits socially and emotionally for the kids involved.[19] Louv and others have been especially radical in suggesting that allowing kids to roam free, out of the sight of hovering parents, to explore, find, discover, and play, is an essential ingredient of childhood, necessary as the foundations that lead to resilient, sensitive, and considerate adults. The essential element, these experts argue, is letting this occur outside, in the open air, in direct contact with nature. What is astounding, profoundly sad as well as inspiring with all this is that these dramatic new interventions are actually necessary in the first place. They appear to be attempting to re-create what were probably the normal childhood experiences that many of us experienced before the advent of the digital age. Such are the conditions we find in the era of the Anthropocene. Radical solutions that reconnect us with nature.

What all these earnest and essential studies and programs seem to be telling us is that the world is in genuine need of simple, easy, local, safe, and personally appropriate ways to engage with the natural world. The benefits for physical and mental health and overall well-being, as well as positive personal interactions and relationships, may all be healthy outcomes. At least as important, there is growing evidence that people who feel more connected to nature—in a wide variety of ways—also tend to be more interested in the wider world and the state of the environment and more willing to be engaged in positive change. These studies also suggest that the surest way to appropriate these benefits is through direct contact. Deliberate, hands-on, dirty fingernails, try-it-and-see-what-happens contact.

In this light, your humble bird feeder takes on a new glow of relevance. It may simply be a way to attract nice birds or it may have a role in saving the world. Whatever its place in your life, it is most certainly more than just a place to see birds. Your feeder is one link in a gigantic chain, a strand in an enormous web, a node in a global communication network. Your private, personal action of providing food for birds changes the structure of an entire, interconnected ecosystem. Your decision to help may alter the dynamics of the evolutionary process and may assist in the process of natural selection in the form of facilitating the spread of diseases. Your feeder is connected, ultimately, to my feeder. My practices, in turn, will affect what happens at your place, eventually.

We think our feeders are for the birds. Our feeders are actually for us. But the birds don't seem to mind. They continue to willingly bring their lives into ours by visiting, and so offer us wonder, hope, knowledge, and pleasure.

Appendix

Species Mentioned in the Text

Common name used in geographical context	**Scientific name**
Plants	
Kauri	Agathis australis
Pink Pine	Halocarpus biformis
Pōhutukawa	*Metrosideros excelsa*
Rimu	Dacrydium cupressinum
Birds	
Adzebill (North Island)	*Aptornis otidiformes*
American Goldfinch	*Spinus tristis*
American Robin	*Turdus migratorius*
American Tree Sparrow	*Spizelloides arborea*
American Wood Duck	*Aix sponsa*
Anna's Hummingbird	*Calypte anna*

Australian Magpie	*Cracticus tibicen*
Bar-headed Goose	*Anser indicus*
Bellbird (New Zealand)	*Anthornis melanura*
Blackbird (Common)	*Turdus merula*
Blackcap	*Sylvia atricapilla*
Black-capped Chickadee	*Poecile atricapillus*
Black Woodpecker	*Dryocopus martius*
Blue Tit	*Cyanistes caeruleus*
Bower's Shrike-Thrush	*Colluricincla boweri*
Brambling	*Fringilla montifringilla*
Bullfinch	*Pyrrhula pyrrhula*
Bush Wren	*Xenicus longipes*
Brehm's Tiger Parrot	*Psittacella brehmii*
Bridled Honeyeater	*Lichenostomus frenatus*
Buzzard (Common)	*Buteo buteo*
Canada Goose	*Branta canadensis*
Carolina Wren	*Thryothorus ludovicianus*
Carrion Crow	*Corvus corone*
Chaffinch	*Fringilla coelebs*
Coal Tit	*Periparus ater*
Cockatiel	*Nymphicus hollandicus*
Collared Dove	*Streptopelia decaocto*
Common (Indian) Myna	*Sturnus tristis*
Common Redpoll	*Carduelis flammea*
Crane (Common)	*Grus grus*
Dunnock	*Prunella modularis*
Eastern Phoebe	*Sayornis phoebe*
Eider (Common)	*Somateria mollissima*
Emerald Dove	*Chalcophaps indica*
Figbird	*Sphecotheres viridis*
Fireback Pheasant	*Lophura ignita*
Florida Scrub Jay	*Aphelocoma caerulescens*
Galah	*Eolophus roseicapilla*
Goldcrest	*Regulus regulus*
Golden-crowned Kinglet	*Regulus satrapa*
Goldfinch (European)	*Carduelis carduelis*
Great Grey Shrike	*Lanius meridionalis*
Great Spotted Woodpecker	*Dendrocopos major*
Great Tit	*Parus major*

Greenfinch	*Carduelis chloris*
Gray Butcherbird	*Cracticus torquatus*
Greylag Goose	*Anser anser*
Grey Warbler	*Gerygone igata*
Haast's Eagle	*Aquila moorei*
Hihi (Stitchbird)	*Notiomystis cincta*
Hooded Merganser	*Lophodytes cucullatus*
House Finch	*Haemorhous mexicanus*
House Sparrow	*Passer domesticus*
Hutton's (Chatham Island) Rail	*Cabalus modestus*
Indigo Bunting	*Passerina cyanea*
Jackdaw	*Corvus monedula*
Kākā	*Nestor meridionalis*
Kakapo	*Strigops habroptilus*
Kea	*Nestor notabilis*
Kererū	*Hemiphaga novaeseelandiae*
Kestrel (Common)	*Falco tinnunculus*
Kōkako (South Island)	*Callaeas cinerea*
Laughing Kookaburra	*Dacelo novaeguineae*
Laughing Owl	*Sceloglaux albifacies*
Macleay's Honeyeater	*Xanthotis flaviventer*
Mallard	*Anas platyrhynchos*
Marsh Tit	*Poecile palustris*
Mauritius Fody	*Foudia rubra*
Mauritius Kestrel	*Falco punctatus*
Mauritius Parakeet	*Psittacula eques*
Mistle Thrush	*Turdus viscivorus*
Moho	*Porphyrio mantelli*
Monk Parakeet	*Myiopsitta monachus*
Namaqua Dove	*Oena capensis*
Noisy Miner	*Manorina melanocephala*
Northern Cardinal	*Cardinalis cardinalis*
Northern Mockingbird	*Mimus polyglottos*
North Island Snipe	*Coenocorypha barrierensis*
Peregrine Falcon	*Falco peregrinus*
Pied Butcherbird	*Cracticus nigrogularis*
Pied Currawong	*Strepera graculina*
Pink Pigeon	*Nesoenas mayeri*
Purple Finch	*Haemorhous purpureus*

Rainbow Lorikeet	*Trichoglossus haematodus*
Raven (Common)	*Corvus corax*
Red-backed Shrike	*Lanius collurio*
Red-bellied Woodpecker	*Melanerpes carolinus*
Red-breasted Nuthatch	*Sitta canadensis*
Red-crowned Crane	*Grus japonensis*
Red-crowned Kākāriki (parakeet)	*Cyanoramphus novaezelandiae*
Red Kite	*Milvus milvus*
Redwing	*Turdus iliacus*
Robin (European)	*Erithacus rubecula*
Rock Dove (feral pigeon)	*Columba livia*
Rose-crowned Fruit-Dove	*Ptilinopus regina*
Rose-ringed (Ring-necked) Parakeet	*Psittacula krameri*
Sacred Ibis	*Threskionis aethiopicus*
Seychelles Magpie-Robin	*Copsychus sechellarum*
Sharp-shinned Hawk	*Accipiter striatus*
Silvereye (White-eye, Waxeye)	*Zosterops lateralis*
Siskin (European)	*Spinus spinus*
Snowy Owl	*Bubo scandiacus*
Southern Cassowary	*Casuarius casuarius*
(Spanish) Griffon Vulture	*Gyps fulvus*
Spanish Imperial Eagle	*Aquila adalberti*
Sparrowhawk	*Accipiter nisus*
Spotted Dove	*Streptopelia chinensis*
Spotted Great Rosefinch	*Carpodacus rubicella severtzovi*
Starling (European)	*Sturnus vulgaris*
Sulphur-crested Cockatoo	*Cacatua galerita*
Swamp Sparrow	*Melospiza geogiana*
Takahē (North Island)	*Porphyrio mantelli*
Tīeke (North Island Saddleback)	*Philesturnus rufusater*
Tooth-billed Bowerbird	*Scenopoeetes dentirostris*
Tufted Titmouse	*Baeolophus bicolor*
Tui	*Prosthemadera novaeseelandiae*
Varied Tit	*Sittiparus varius*
Victoria's Riflebird	*Ptiloris victoriae*
Wattled Honeyeater	*Foulehalo carunculatus*
Weka	*Gallirallus australis*
West Peruvian Dove	*Zenaida meloda*

Whinchat	*Saxicola rubetra*
Willow Tit	*Poecile montanus*
Woodpigeon	*Columba palumbus*

Mammals

Common Brushtail Possum	*Trichosurus vulpecular*
Ferret	*Mustela putorius furo*
Hedgehog (European)	*Erinaceus europaeus*
House Mouse	*Mus musculus*
Kiore (Polynesian rat)	*Rattus exulans*
Norway Rat	*Rattus norvegicus*
Polar Bear	*Ursus maritimus*
Ship (Black) Rat	*Rattus rattus*
Stoat	*Mustela erminea*
Weasel	*Mustela nivalis*
Wild boar	*Sus scrofa*

Notes

Preface

1. Reliable figures on participation rates are presented in some detail in Chapter 1, but some key references are Cowie and Hinsley 1988a, Lepczyk et al. 2004, Rollinson et al. 2003, and Davies et al. 2012.

2. Expertly summarized in Baicich et al. 2015 for the United States.

3. See, for example, Orros and Fellowes 2015.

4. See Lin 2005.

1. Why Bird Feeding Matters

1. Strong antifeeding messages can be found in the literature and on the websites of many Australian organizations, including wildlife rescue and rehabilitation groups, bird societies, and environmental agencies. Some typical examples are www.environment.nsw.gov.au/animals/KeepingWildlifeWild.htm and www.abc.net.au/news/2016-01-27/why-you-should-not-feed-native-birds/7118192. Typical titles include "Why We Don't Feed the Wildlife," "If You Care about Wildlife," "Let Nature Feed Itself," "Recipe for Disaster: Feed the Animals," and "Remember, You're Killing Them with Kindness." See Jones 2011 for an extensive list of relevant websites.

2. These ideas are explored vividly by Tim Low in *Where Song Began* (2014).

3. See Plant 2008 and Parsons 2008 for discussion of "planting for birds" in Australia.

4. Tim Low, again, describes this phenomenon expertly in *Where Song Began* as well as in *The New Nature* (2002). For more formal evidence, see French et al. 2004 and Parsons et al. 2006.

5. Carla Catterall (2004) has summarized the unexpected interactions of garden plantings and the birds they attract.

6. The dramatic increase in the abundance of Rainbow Lorikeets and the ecological implications are described in Smith and Lill 2008, Davis et al. 2011, and Jaggard et al. 2015.

7. See Low 2014.

8. See Low 2002.

9. The diet of urban Australian Magpies is described in Rollinson and Jones 2003, O'Leary and Jones 2006, Ishigame and Baxter 2006, and summarized in Jones 2002.

10. See Jones 2002 for the full story of the Australian Magpie as an urban success.

11. See Jones 2002.

12. The characteristics of species that have successfully invaded the suburban environment are explained and discussed in John Marzluff's *Subirdia* (2014); more technical discussions can be found in Marzluff et al. 2001 and Chace and Walsh 2006.

13. See details mentioned in Note 1.

14. See Jones and Reynolds 2008.

15. There are many such sources, but some well-known examples are the websites of the British Trust for Ornithology, Cornell Lab of Ornithology, and Royal Society for the Protection of Birds. Search for "bird feeding."

16. This is the unequivocal stance of almost all North American sources and exemplified by the new book *Feeding Wild Birds in America* (Baicich, Barker, and Henderson 2015), though Sterba 2012 presents a contrary view.

17. See, for example, Deis 1982, Bird 1986, and Kobilinsky 2015.

18. This claim was first made, as far as I can tell, in Jones and Reynolds 2008.

19. See Cowie and Hinsley 1988a, b.

20. Memorably described in Eric Rolls's classic *They All Ran Wild* (1969).

21. Published as Jones 1981.

22. See Thomas 2000.

23. See Cannon 1999, 288.

24. See McLees 2001.

25. See Howard 2006, Howard and Jones 2004, and Rollinson et al. 2003.

26. See Ishigame and Baxter 2007.

27. Although the link to this project by the Australian Broadcasting Commission is no longer operating, a description of its aims and procedures can be found at http://www.abc.net.au/radionational/programs/scienceshow/wildwatch-australia/3374512.

28. The New Zealand Garden Bird Survey website is at http://www.landcareresearch.co.nz/science/plants-animals-fungi/animals/birds/garden-bird-surveys.

29. Visit BTO's Garden BirdWatch at www.bto.org.gbw

30. For details, see Galbraith et al. 2014.

31. See Orros and Fellowes 2015.

32. A component of the study described in Fuller et al. 2008.

33. See Lepczyk et al. 2012.

34. The US Fish and Wildlife Service publishes the National Survey of Fishing, Hunting, and Wildlife-Associated Recreation. Recent surveys are available at http://wsfrprograms.fws.gov/.

35. See Note 34.

36. Presented formally in the Department of the Environment, Food and Rural Affairs (DEFRA) report *Working with the Grain of Nature* (2002) and reassessed by Fuller et al. 2008.

37. Data derived from the CityForm surveys were utilized by a consortium of researchers (see Gaston et al. 2007); the relevant information reported here is from Davies et al. 2012.

38. See Davies et al. 2012.

39. Similarly described in Gaston et al. 2007 and analyzed by Davies et al. 2012.

40. See, for example, Baicich et al. 2015.

41. See estimates in Baicich et al. 2015, US Fish and Wildlife Service 2012 for the United States, and the Horticultural Trades Association website for the UK, www.the-hta.org.uk/page.php?pageid=187.

42. Data from US Fish and Wildlife Service 2012, Baicich et al. 2015, and Glue 2006.

43. See Galbraith et al. 2014.

44. See Robb et al. 2008a.

45. See Orros and Fellowes 2015.

46. See websites listed in Jones 2011.

47. Material used for the original article (Sterba 2002) was recycled in the book *Nature Wars* (Sterba 2015) with, disappointingly, virtually no alteration of opinion or inclusion of new information.

48. Refer to details in Jones 2011.

2. Crumbs to Corporations

1. Figures supplied directly by Chris Whittles in November 2014.

2. A cursory search turns up a plethora of British suppliers: Arkwildlife, British Bird Seed, Colonels, Food4WildBirds, Garden Wildlife Direct, Gardmans, Haiths, Kennedy, Really Wild, Vinehouse, Wild Bird Feed. Amazon and eBay now offer a variety of products.

3. Cited in Fuller et al. 2012.

4. Some of these companies are mentioned in Baicich et al. 2015, but each has its own informative website describing their history.

5. The comprehensive US Fish and Wildlife surveys are an invaluable source of detailed and reliable information of relevance here; the surveys are available for free from the agency's website (www.fws.gov).

6. See report by Lin 2005 from the United Nations Food and Agriculture Organisation.

7. See Lin 2005.

8. This information comes from Baicich, Barker, and Henderson's booklet *Feeding Wild Birds* (2010), the precursor to their much more detailed book *Feeding Wild Birds in America* (2015).

9. I visited Mr. Whittles at his home in November 2014.

10. See details in note 8.

11. These studies were published as Chapman and Jones 2009, 2010, 2012.

12. See Cocker and Tipling 2014.

13. Read more about the Pancha-Maha-yajnas at http://veda.wikidot.com/panchamahayajna.

14. These details are from Bailleul-LeSuer 2014.

15. See Bailleul-LeSuer 2014.

16. See Wade et al. 2012.

17. See Wade et al. 2012.

18. These accounts were sourced from standard websites, though some details are also based on David Elliston Allen's excellent *The Naturalist in Britain: A Social History* (1976).

19. See Augustine Thompson's *Francis of Assisi: A New Biography* (2012).

20. *Northampton Mercury*, 28 April 1787, under "Country News."

21. "Trial of McNaughton for the murder of Mr. Drummond, " *Caledonian Mercury*, 9 March 1843.

22. See Thoreau (1854) 1946.

23. For example, see Baicich et al. 2010.

24. See Baicich et al. 2015.

25. See Baicich et al. 2015.

26. See Renee Thompson's *The Plume Hunter* (2011).

27. See Baicich et al. 2015.

28. Apparently first described by Dovaston in the *Magazine of Natural History* in 1832 and discussed in Allen 1976.

29. See Allen 1976.

30. See Callahan 2014.

31. See Callahan 2014.

32. *Ipswich Journal*, 3 February 1776, under "Colchester."

33. "Miscellaneous reflections on the study of nature," *Kendal Mercury*, 5 May 1855.

34. "Feeding birds in winter," *Edinburgh Evening News*, 16 February 1875.

35. See Callahan 2014.

36. See Callahan 2014.

37. Two articles in *London Daily News*, 30 December 1890: "The birds," and "A thrush on the 'situation.'"

38. See Callahan 2014.

39. See note 37.

40. See note 37.

41. See Allen 1976.

42. See Allen 1976.

43. *How to Attract and Protect Wild Birds* was written by Martin Hiesemann and translated by Emma S. Buchheim. The Duchess of Bedford wrote the introduction. It contains many illustrations.

44. The baron's book, originally published in 1899, was *Der gesamte Vogelschutz: Seine Begründung und Ausführung* (Gera-Untermhaus, Köhler).

45. See Hiesemann 1908.

46. See Hiesemann 1908.

47. See McAtee 1914.

48. See Forbush 1918.

49. Forbush 1918.

50. An excellent exploration of the philosophy and significance of national parks is *National Parks beyond the Nation* (2012), edited by Howkins et al.

51. See Howkins et al. 2012.

52. Cited by Soper 1965.

53. See Baicich et al. 2015.

54. See Baicich et al. 2015.

55. See Baicich et al. 2015.

56. Cited by Baicich et al. 2015.

57. See Barker and Griggs 2000 and Baicich et al. 2015.

58. See Baicich et al. 2010.

59. See Baicich et al. 2010.

60. See Callahan 2014.

61. See Baicich et al. 2015.

62. See Baicich et al. 2010.

63. Baicich et al. 2015. See also McCormick et al. 1992.

64. See McCormick et al. 1992.

65. See McCormick et al. 1992.

66. See Baicich et al. 2010.

67. See Baicich et al. 2015.

68. See Baicich et al. 2015.

69. My master's thesis was entitled "Parrots and sunflowers in northern New South Wales" (University of New England, Armidale, 1982).

70. See Baicich et al. 2010. The spellings "nyjer" and "nyger" are both widely used; I have chosen "nyger" for this book.

71. T. J. Greenaway 1988.

72. See Geis 1980.

73. See Baicich et al. 2010.

74. See Baicich et al. 2010.

75. See note 9.

76. There is active debate about wild bird feeding in South Africa. For some of the range of opinions, visit http://www.motherearthnews.com/nature-and-environment/feeding-wild-birds-zm0z92zros.aspx and http://www.birdlife.org.za/about-us/frequently-asked-questions.

3. The Big Change

1. See Berthold and Mohr 2006.

2. For a sampling of opposing reviews, in German, visit www.bund.net/ueber_uns.

3. See Baicich et al. 2010, 2015.

4. See Belaire et al., 2015, Lepczyk et al. 2004, Lepczyk et al. 2012.

5. See Baicich et al. 2010.

6. See Lepczyk et al. 2004.

7. See Lepczyk et al. 2004.

8. This was certainly the strong assumption of Tony Soper in *The Bird Table Book* (1965), the "bible of garden birding."

9. See Cowie and Hinsley 1988a, b.

10. See Davies et al. 2012.

11. See Orros and Fellowes 2015.

12. Quoted by Thompson 1987, referring to RSPB leaflets "Food fit for birds?" and "Feeding garden birds."

13. The earlier advice from BTO was largely distributed as leaflets that are no longer available, although the general attitude is repeated in the popular books of the time such as Tony Soper's *Bird Table Book* (1965) and Glue's *Garden Bird Book* (1982).

14. See, for example, Toms and Sterry's *Garden Birds and Wildlife* (BTO, 2008).

15. Current views from RSPB can be found at http://www.rspb.org.uk/makeahomefor wildlife/advice/helpingbirds/feeding.

16. See Plummer et al. 2013.

17. See Thompson 1987.

18. See, for example, Cowie and Hinsley 1987, 1988b.

19. See discussion in Cannon 1999 and Chamberlain et al. 2009.

20. Learn about BTO Garden BirdWatch at www.bto.org.gbw.

21. And to prove that the "Oddie Effect" is alive and well, check out the collaboration between Bill and the seed company Haiths: "It's time to join Bill's *Summer Feeding is Cool!* revolution." http://www.haiths.com/its-time-to-join-bill-oddies-summer-bird-feeding-is-cool-revolution.

22. See, for example, www.bund.net/ueber_uns.

23. See, for example, www.bund.net/ueber_uns.

24. See, for example, Berthold et al. 1992, Berthold 1996.

25. See Berthold and Mohr 2006.

26. See Berthold and Mohr 2006.

27. See Heisemann 1908.

28. See Berthold and Mohr 2006.

4. The Feeder Effect

1. There is an abundance of information on Project FeederWatch on their website, FeederWatch.org, or via the Lab of Ornithology, www.birds.cornell.edu.

2. Visit www.birds.cornell.edu.

3. Just a couple of the many excellent citizen science publications from the Lab of O are Cooper et al. 2007 and Bonter and Cooper 2012.

4. A useful summary of the history and development of Project FeederWatch is provided by Barker and Grigg 2000.

5. See Barker and Grigg 2000.

6. See Barker and Grigg 2000.

7. This is easy. Visit FeederWatch.org/pfw/maps and choose your favorite species.

8. See Leston and Rodewald 2006.

9. See FeederWatch.org/pfw/maps.

10. The extraordinary story has been led and communicated by the Cornell Lab of Ornithology's Andre Dhondt from the very beginning. See, for example, Bonney and Dhondt 1997, Dhondt et al. 1995, and Dhondt et al. 2005.

11. Scc Dhondt et al. 1995.

12. See Fischer and Miller 1995.

13. See Adelman et al. 2015.

14. See Fischer and Miller 2015.

15. Visit the National Audubon Society's Christmas Bird Count site: www.christmasbirdcount.org.

16. See Fischer and Miller 2015.

17. Visit BTO Garden BirdWatch at www.bto.org.gbw.

18. See Berthold et al. 1992.

19. See Bearhop et al. 2005.

20. See Bearhop et al. 2005.

21. See Plummer et al. 2015.

22. See Plummer et al. 2015.

23. See Bearhop et al. 2005.

24. See Rolshausen et al. 2009.

25. See Brittingham and Temple 1988 and Brittingham 1991.

26. See, for example, Kluyver 1952.

27. Talvised Aialinnud, Ja Nende Toitmine (2011). Eesti Ornitoloogiaühing, Tartu, Estonia. Visit www.eoy.ee.

28. See Thompson 1987.

29. Quoted in Thompson 1987.

30. For example, see Howard and Jones 2004 and Jones and Reynolds 2008.

31. See Jones and Reynolds 2008.

32. See, for example, Soper 1965 and Peterson 2000.

33. See Brittingham and Temple 1992a.

34. See Brittingham and Temple 1992a.

35. Wilcoxen et al. 2015. Their project remains one of the most important studies on the effects of feeding.

36. See Galbraith et al. 2015.

37. Summarized in Jones 2002.

38. See O'Leary and Jones 2006.

39. See Cowie and Hinsley 1988b.

40. Discussed in Jones 2011.

41. See, for example, Thompson 1987.
42. See Cowie and Hinsley 1988b.
43. See Schoech and Bowman 2003.
44. See Peach 2015.
45. See Chamberlain et al. 2009 and Orros and Fellowes 2015.
46. See Lepczyk et al. 2004.
47. See Fuller et al. 2012.
48. See Fuller et al. 2008 and Fuller et al. 2012.
49. See Wilcoxen et al. 2015.
50. See Fuller et al. 2008 and Fuller et al. 2012.
51. See Lepzyyk et al. 2004.
52. John Marzluff's *Welcome to Subirdia* (2014) summarizes this expertly.
53. See Chamberlain et al. 2009.
54. See Fuller et al. 2008.
55. See Galbraith et al. 2014.
56. Observations by Chris Whittles but widely noted (see Toms and Sterry 2008).
57. Visit BTO Garden BirdWatch at www.bto.org.gbw.
58. See Orros and Fellowes 2014.
59. See Galbraith et al. 2014.
60. See South and Pruett-Jones 2000.
61. See Peck et al. 2014.
62. See Peck et al. 2014.

5. What Happens When We Feed?

1. A summary of our research on Australian Magpies and especially their aggression toward humans is provided in Jones 2002.
2. See Jones 2002.
3. See Jones 2002.
4. This surprising finding was reported in Hughes et al. 2003.
5. Details provided in Jones 2002.
6. Floyd and Woodland 1981.
7. Our findings are presented in O'Leary and Jones 2006.
8. These studies are summarized in Jones 2002.
9. See Boutin 1990.
10. See Watson 1970.
11. This is described in detail by Dhondt et al. 1984.
12. See Clamens and Isenmann 1989.
13. An excellent synthesis of this large body of work is provided by Robb et al. 2008a.
14. See Svensson and Nilsson 1995.
15. Two foundational papers on these topics are Drent and Daan 1980 and Meijer and Drent 1999. Genuinely essential readings.
16. A key reference for this topic is Robb et al. 2008a.
17. This important study was published as Robb et al. 2008b.
18. See Robb et a. 2008b.
19. See Harrison et al. 2010.
20. See Harrison et al. 2010.
21. See Hill 1988.
22. See Robb et al. 2008a.
23. See Chamberlain et al. 2009.

24. This key reference is Chamberlain et al. 2009.
25. See Plummer et al. 2013.
26. See Plummer et al. 2013.
27. See Plummer et al. 2015.
28. The standard reference for all aspects of nutrition in wildlife generally is still Robbins 1983.
29. See Plummer et al. 2013.
30. See Robb et al. 2008b.
31. See Plummer et al. 2013.
32. See Kallander 1981.
33. Kallander 1981, 247.
34. See Smith 1991.
35. See Smith 1967.
36. See Brittingham and Temple 1988.
37. See Brittingham and Temple 1988.
38. See Brittingham and Temple 1988, 587.
39. See Brittingham and Temple 1988, 587.
40. See Jansson et al. 1981.
41. See Robb et al. 2008a.
42. See Wilson 2001.
43. See Grubb 1987.
44. See Kubota and Nakamura 2000.
45. See Saggese et al. 2011.
46. See Poesel et al. 2006.
47. See Otter et al. 1997.
48. See Cuthill and Macdonald 1990.
49. See Barnett and Briskie 2007.
50. See Saggese et al. 2011.
51. See Kacelnik 1979.
52. See Woolfenden and Fitzpatrick 1984.
53. A comprehensive review of the issues is given in Hatchwell 2009.
54. See Schoech 1996.
55. See Schoech 1996, 236.
56. See Fleischer et al. 2003.
57. See Schoech and Bowman 2003.
58. See Reynolds et al. 2003.
59. See Ishigame et al. 2006.
60. See Boutin 1990.
61. Robb et al. 2008b.

6. Tainted Table?

1. As described in Margaret Barker and Jack Griggs's 2000 book, *The FeederWatcher's Guide to Bird Feeding* (sponsored by the Cornell Lab of Ornithology).
2. See Fischer et al. 1997.
3. See Ley et al. 1996.
4. The whole story is described in detail in Dhondt et al. 2005.
5. One of the first reports from this group is Dhondt et al. 1995.
6. See Dhondt et al. 2005.
7. See Hosseini et al. 2004.

8. See Dhondt et al. 2007a.
9. See Adelman et al. 2015.
10. See Adelman et al. 2015.
11. See details in Dhondt et al. 2007b.
12. See Adelman et al. 2015.
13. See Adelman et al. 2015.
14. See Hofle et al. 2004.
15. See Hofle et al. 2004.
16. See Hanson 1969.
17. See Hofle et al. 2004.
18. Robinson et al. (2010) provides an excellent, detailed account of this outbreak and its impact on birds in the UK.
19. See Robinson et al. 2010.
20. See Robinson et al. 2010.
21. The Garden Bird Health Initiative is hosted by the Zoological Society of London (visit www.zsl.org/science/research/gwh.
22. See Robinson et al. 2010.
23. See Robinson et al. 2010.
24. See Lawson et al. 2012.
25. See Ganas et al. 2014.
26. See Ganas et al. 2014.
27. Mentioned in Ganas et al. 2010.
28. See Robinson et al. 2010.
29. See Robsinson et al. 2010.
30. Described more generally by Becker et al. 2015.
31. See Lawson et al. 2012.
32. See Lawson et al. 2012.
33. See Lawson et al. 2012.
34. Quotation from Lawson et al. 2012, 7.
35. See Tollington et al. 2015.
36. The leaflet "Beak and Feather Disease *(Psittacine ciroviral disease)*" is available from the Australian Department of the Environment and Heritage (www.deh.gov.au).
37. A major source of relevant information is Thomas et al. 2007.
38. Mentioned in Lawson et al. 2012.
39. See Adelman et al. 2015.
40. Mentioned by Becker et al. 2015.
41. See Dhondt et al. 2007a.
42. Chris Whittles, pers. comm., October 2015.
43. Cornell University, Department of Animal Science, http://poisonousplants.ansci.cornell.edu/toxicagents/aflatoxin.
44. See note 43
45. See Lawson et al. 2006.
46. See Leung et al. 2006.
47. Visit the Birdcare Standards Association at www.birdcare.org.uk.
48. See note 43.
49. See Robinson et al. 2010.
50. See note 43.
51. See Robinson et al. 2010.
52. GrrlScientist 2012.
53. GrrlScientist 2012.

54. The standard reference for all aspects of nutrition in wildlife generally is still Robbins 1983, though Barboza et al. 2009 provides some important new material.

55. See Frith and Frith 2001.

56. See note 54.

57. See note 47.

58. See note 47.

59. Description provided by Sterba 2012.

60. See Sterba 2012.

61. See Waters et al. 2009.

62. Among the exceptions are several studies led by Renee Chapman in Australia. See Chapman and Jones 2009, 2010, 2012.

63. Visit the BTO and Cornell Lab websites for various opinions on the use of bread today, noting that it remains extremely widespread as a food offered to birds throughout the world.

64. See Rollinson and Jones 2003.

65. See Ishigame et al. 2006.

66. See Robbins 1983.

67. See Orros and Fellowes 2014.

68. Described in Jones 2002 and O'Leary and Jones 2006.

69. Just Google "meat-eating lorikeets" and stand back!

7. Feeding for a Purpose

1. The dramatic and tragic story of the wildlife of New Zealand has been covered by many authors, though Kerry-Jayne Wilson's *Flight of the Huia* (2004) is among the most detailed and reliable.

2. See Galbraith et al. 2014.

3. Tennyson and Martinson 2006.

4. See Tennyson and Martinson 2006.

5. See Worthy and Holdaway 2002.

6. These dates and a host of related issues are discussed in the magisterial *The Lost World of the Moa: Prehistoric Life in New Zealand* by Trevor Worthy and Richard Holdaway (2002).

7. See Worthy and Holdaway 2002.

8. Flannery 2005.

9. See Flannery 2005.

10. See Tennyson and Martinson 2006.

11. See Worthy and Holdaway 2002.

12. Pawson and Brooking 2002.

13. This is an all too familiar scenario, tragically yet brilliantly explained by Stolzenburg 2011.

14. See Tennyson and Martinson 2006.

15. See Stolzenburg 2011.

16. This graphic is found in Tennyson and Martinson 2006.

17. See Stolzenburg 2011.

18. See Stolzenburg 2011.

19. The role and issues associated with the use of supplementary feeding in conservation projects throughout the world are discussed in many places including Blanco et al. 2011, Boutin 1990, and Dubois and Fraser 2013.

20. Visit www.tiritirimatangi.org.nz.

21. Birkhead 2012.
22. See Tennyson and Martinson 2006.
23. See Taylor and Castro 2001.
24. See Castro et al. 2003.
25. See Castro et al. 2003.
26. Taylor and Castro 2001.
27. See Castro et al. 2003.
28. See Castro et al. 2003.
29. See Castro et al. 2003.
30. See Tennyson and Martinson 2006.
31. Again, covered nicely by Stolzenburg 2011.
32. Mentioned in Stolzenburg 2011.
33. See Wilson 2004.
34. See Lloyd and Powlesland 1994.
35. See Wilson 2004.
36. See Stolzenburg 2011.
37. See Stolzenburg 2011.
38. See Lloyd and Powlesland 1994.
39. See Lloyd and Powlesland 1994.
40. See James et al.1991 and Powlesland and Lloyd 1994.
41. See Powlesland and Lloyd 1994.
42. See Clout et al. 2002.
43. See Lloyd and Powlesland 1994.
44. See Lloyd and Powlesland 1994.
45. See Powlesland and Lloyd 1994.
46. See Powlesland and Lloyd 1994.
47. See Powlesland and Lloyd 1994.
48. See Powlesland and Lloyd 1994.
49. See Powlesland and Lloyd 1994.
50. See Powlesland and Lloyd 1994.
51. Ralph Powlesland personal communication 2015.
52. See Powlesland and Lloyd 1994.
53. See Powlesland and Lloyd 1994.
54. See Clout et al. 2002.
55. See Clout et al. 2002.
56. See Tella 2001.
57. See Houston et al. 2007.
58. Quotation from Clout et al. 2002, 17.
59. See Clout et al. 2002.
60. See Edmunds 2008.
61. The standard general reference is still Martin 1987.
62. See Wright and Leonard 2007.
63. See Martin et al. 2000.
64. See Vickery et al. 2004.
65. Recorded by Charles Swainson (1885) 2004.
66. Deftly described in *Sparrow* by Kim Todd (2012).
67. See Todd 2012.
68. Jones 1981, 1983.
69. See Summers-Smith 2003.
70. See Chamberlain et al. 2007.

71. See Chamberlain et al. 2007.
72. See Peach et al. 2008.
73. Summarized in detail in Baillie et al. 2007.
74. See Peach et al. 2008.
75. See Peach et al. 2008.
76. See Chamberlain et al. 2007.
77. See Peach et al. 2008.
78. See Peach et al. 2008.
79. See Peach et al. 2015.
80. See Peach et al. 2015.
81. Information from Will Peach, personal communication, 2014.
82. See Peach et al. 2014.
83. Shakespeare quoted in McCarthy 2006.
84. Ackroyd 2000.
85. The standard account is Carter 2007.
86. See Carter 2007.
87. See Carter 2007.
88. An excellent summary of the Red Kite story can be found at the RSPB site: www.rspb.org.uk/discoverandenjoynature/discoverandlearn/birdguide/name/r/redkite/.
89. See McCarthy 2006.
90. See http://www.gigrin.co.uk.
91. More details are available at http://www.chilternsaonb.org/about-chilterns/red-kites.html.
92. The Gigrin Farm's history and current details are found at http://www.gigrin.co.uk.
93. See Carter 2007.
94. See http://www.gigrin.co.uk.
95. See Carter 2007.
96. See Orros and Fellowes 2014.
97. An article questioning the previously promoted practice of feeding red kites is available at http://web.onetel.com/~gerrywhitlow/sekg/kite%20feeding%20guidelines.pdf.
98. See Orros and Fellowes 2014.
99. Comments by Melanir Orros, personal communication, 2014.
100. See Orros and Fellowes 2014.
101. Ecological traps in urban environments are nicely discussed by Leston and Rodewald 2006.
102. See Leston and Rodewald 2006.
103. See Parra and Tellería 2004.

8. Reasons Why We Feed Wild Birds

1. One part of Peter Howard's PhD thesis was published as Howard and Jones 2004.
2. Some of these issues are described in Nattrass 2001, Jones and Howard 2001, Jones 2011, and Jones 2014.
3. See Howard and Jones 2001.
4. See, for example, Clucas and Marzluff 2012 and Clucas et al. 2014.
5. See Marzluff and Miller 2014.
6. John Marzluff has been investigating corvids for a long time, and his findings are beautifully summarized in *In the Company of Crows and Ravens* (2005) and *Gifts of the Crow* (2012), both written with Tony Angell.

7. See Marzluff and Angell 2005.
8. Schreiber 2010.
9. Clark 2013.
10. See, for example, Fuller and Irving 2010 and Lepczyk et al. 2012.
11. See Chapman and Jones 2009, 2011, 2012.
12. Chapman 2015.
13. Kellert 1996, 1997.
14. One of the most thoughtful and wide ranging explorations of the relationships between people and wildlife is Michael Manfredo's *Who Cares about Wildlife* (2008), and this includes excellent discussion of Kellert's values system.
15. See Miller 2005.
16. Discussed critically in Perkins 2010.
17. See note 16.
18. See Cox and Gaston 2016.
19. See Cowie and Hinsley 1988a.
20. Cox and Gaston 2016, 10–11.

9. Bird Feeding Matters Even More Now

1. Baicich, Barker, and Henderson 2010.
2. An important argument made by Cannon 1999 and Berthold and Mohr 2006.
3. As documented by Baicich et al. 2015.
4. There are plenty of researchers holding such opinions, and Robb et al. 2008a provides a good summary.
5. Coined in Jones and Reynolds 2008.
6. See, for example, Fuller et al. 2008.
7. See Fuller et al. 2008.
8. Or visit Anthropocene.info.
9. See Brittingham and Temple 1992a.
10. See Chace and Walsh 2006.
11. These and related topics are discussed in McDonnell et al. 2009.
12. See Galbraith et al. 2014.
13. See, for example, Ottoni et al. 2009.
14. See Robinson et al. 2010.
15. See Louv 2005.
16. See Louv 2005.
17. Discussed and summarized nicely by Fuller and Irvine 2010.
18. These issues and others are discussed in Selhub and Logan 2012.
19. Visit, for example, www.childrenandnature.org.

References

Ackroyd, P. 2000. *London: The Biography*. London: Nan A. Talese.

Adelman, J. S., S. C. Moyers, D. R. Farine, and D. M. Hawley. 2015. Feeder use predicts both acquisition and transmission of a contagious pathogen in a North American songbird. *Proceedings Royal Society B*, *282*. doi:10.1098/rspb.2015.1429.

Allen, D. E. 1976. *The Naturalist in Britain: A Social History*. Princeton, NJ: Princeton University Press.

Baicich, P. J., M. A. Barker, and C. L. Henderson. 2010. *Feeding Wild Birds: A Short History in America*. Glen Echo, MD: Wild Bird Centers of America.

Baicich, P. J., M. A. Barker, and C. L. Henderson. 2015. *Feeding Wild Birds in America: Culture, Commerce, and Conservation*. College Station: Texas A&M University Press.

Bailleul-LeSuer, R. 2014. *Between Heaven and Earth: Birds in Ancient Egypt*. Chicago: Chicago University Press.

Baillie, S., J. H. Marchant, H. Q. P. Crick, D. G. Noble, D. E. Balmer, C. Barimore, R. Coombes, I. Downie, S. Freeman, A. Joys, D. Leach, M. Raven, R. Robinson, and R. M. Thewlis. 2007. *Breeding Birds in the Wider Countryside: Their Conservation Status 2006*. Thetford: British Trust for Ornithology.

Barbosa, P. S., K. L. Parker, and I. D. Hume. 2009. *Integrative Wildlife Nutrition*. Berlin: Springer-Verleg.

Barker, M. A., and J. Griggs. 2000. *The FeederWatcher's Guide to Bird Feeding*. New York: HarperCollins.

Barnett, C. A., and Briskie, J. V. 2007. Energetic state and the performance of dawn chorus in silvereyes (*Zosterops lateralis*). *Behavioral Ecology and Sociobiology* 61:579–587.

Bearhop, S., W. Fiedler, R. W. Furness, S. C. Votier, J. Newton, G. J. Bowen, P. Bethold, and K. Farnswortha. 2005. Assortative mating as a mechanism for rapid evolution of a migratory divide. *Science* 310(5747): 502–504.

Becker, D. J., D. G. Streicker, and S. Altizer. 2015. Linking anthropogenic resources to wildlife-pathogen dynamics: A review and meta-analysis. *Ecology Letters* 18(5): 483–495.

Berthold, P. 1996. *Control of Bird Migration*. London: Chapman and Hall.

Berthold, P., A. J. Helbig, G. Mohr, and U. Querner. 1992. Rapid microevolution of migratory behaviour in a wild bird species. *Nature* 360:668–670.

Berthold, P., and G. Mohr. 2006. *Vögel füttern, aber richtig*. Stuttgart: Kosmos.

Bird, D. M. 1986. City critters: A cast of millions. *In* L. W. Adams and D. L. Leedy, eds., *Integrating Man and Nature in the Metropolitan Environment*, pp. 23–27. Chevy Chase, MD: National Institute for Urban Ecology.

Birkhead, T. 2012. *Bird Sense: What It's Like to Be a Bird*. London: Bloomsbury.

Blanco, G., J. A. Lemus, and M. Garcia-Montijano. 2011. When conservation management becomes contraindicated: Impact of food supplementation on health of endangered wildlife. *Ecological Applications* 21:2469–2477.

Bonney, R., and A. A. Dhondt. 1997. FeederWatch: An example of a student-scientist partnership. In K. C. Cohen, ed., *Internet Links for Science Education*, pp. 31–53. New York: Plenum Press.

Bonter, D. N., and C. B. Cooper. 2012. Data validation in citizen science: A case study from Project FeederWatch. *Frontiers in Ecology and the Environment 10*(6): 305–307.

Boutin, S. 1990. Food supplementation experiments with terrestrial vertebrates: Patterns, problems, and the future. *Canadian Journal of Zoology* 68(2): 203–220.

Brittingham, M. C. 1991. Effects of winter bird feeding on wild birds. *In* L. W. Adams and D. L. Leedy, eds., *Conservation in Metropolitan Environments*, pp. 185–190. Columbia, MD: National Institute for Urban Wildlife.

Brittingham, M. C., and S. A. Temple. 1988. Impacts of supplementary feeding on survival rates of black-capped chickadees. *Ecology* 69:581–589.

Brittingham, M. C., and S. A. Temple. 1992a. Does winter bird feeding promote dependency? *Journal of Field Ornithology* 63(2): 190–194.

Brittingham, M. C., and S. A. Temple. 1992b. Use of winter bird feeders by black-capped chickadees. *Journal of Wildlife Management* 56:103–110.

Callahan, D. 2014. *A History of Birdwatching in 100 Objects*. London: Bloomsbury.

Cannon, A. 1999. The significance of private gardens for bird conservation. *Bird Conservation International* 9:287–297.

Carter, I. 2007. *The Red Kite*. Shrewsbury: Arlequin Press.

Castro, I. 1995. Behavioural ecology and management of hihi Notiomystis cincta, an endemic New Zealand honeyeater. PhD thesis, Massey University, New Zealand.

Castro, I., D. H. Brunton, K. M. Mason, B. Ebert, and R. Griffiths. 2003. Life history traits and food supplementation affect productivity in a translocated population of the endangered Hihi (Stitchbird, *Notiomystis cincta*). *Biological Conservation* 114(2): 271–280.

Castro, I., K. M. Mason, D. P. Armstrong, and D. M. Lambert. 2004. Effect of extra-pair paternity on effective population size in a reintroduced population of the endangered hihi, and potential for behavioural management. *Conservation Genetics* 5:381–393.

Catterall, C. P. 2004. Birds, garden plants and suburban bushlots: Where good intentions meet unexpected outcomes. *In* S. Burgin and D. Lunney, eds., *Urban Wildlife: More Than Meets the Eye*, pp. 21–31. Sydney: Royal Zoological Society of NSW.

Chace, J. F., and J. J. Walsh. 2006. Urban effects on native avifauna: A review. *Landscape and Urban Planning* 74:46–69.

Chamberlain, D. E., A. R. Cannon, M. P. Toms, D. I. Leech, B. J. Hatchwell, and K. J. Gaston. 2009. Avian productivity in urban landscapes: A review and meta-analysis. *Ibis* 151(1): 1–18.

Chamberlain, D. E., M. P. Toms, R. Cleary-McHarg, and A. N. Banks. 2007. House sparrow (*Passer domesticus*) habitat use in urbanized landscapes. *Journal of Ornithology* 148:453–462.

Chapman, R. A. 2015. Why do people feed wildlife? An international comparison. PhD thesis, Griffith University, Brisbane, Australia.

Chapman, R., and D. Jones. 2009. Just feeding the ducks: Quantifying a common wildlife-human interaction. *Sunbird* 39:18–28.

Chapman, R., and D. Jones. 2010. Duck diversity in Greater Brisbane: Native species, domestic races and the influence of feeding. *Sunbird* 40:29–38.

Chapman, R., and D. Jones. 2012. Synurbisation of Pacific Black Ducks *Anas superciliosa* in south-eastern Queensland: The influence of supplementary feeding on foraging behaviour. *Australian Field Ornithology* 29:31–39.

Clamens, A., and P. Isenmann. 1989. Effects of supplemental food on the breeding of Blue and Great Tits in Mediterranean habitats. *Ornis Scandinavica* 20:36–42.

Clark, D. 2013. A Study of the Motivation of the General Public in Feeding Birds in Their Gardens. MSc in Ornithology, Birmingham University, Birmingham, U.K.

Clout, M. N., G. P. Elliot, and B. C. Robertson. 2002. Effects of supplementary feeding on the offspring sex ratio of kakapo: A dilemma for the conservation of a polygynous parrot. *Biological Conservation* 107:13–18.

Clucas, B., and J. M. Marzluff. 2012. Attitudes and actions toward birds in urban areas: Human cultural differences influence bird behavior. *Auk* 129:8–16.

Clucas, B., S. Rabotyagov, and J. M. Marzluff. 2014. How much is that birdie in my backyard? A cross-continental economic valuation of native urban songbirds. *Urban Ecosystems* 18:251–266.

Cocker, M., and D. Tipling. 2014. *Birds and People*. London: Random House.

Cooper, C. B., J. Dickinson, T. Phillips, and R. Bonney. 2007. Citizen science as a tool for conservation in residential ecosystems. *Ecology and Society* 12(2):1–12.

Cowie, R. J., and S. A. Hinsley. 1987. Breeding success of Blue Tits and Great Tits in suburban gardens. *Ardea* 75:81–90.

Cowie, R. J., and S. A. Hinsley. 1988a. Feeding ecology of Great Tits (*Parus major*) and Blue Tits (*Parus caeruleus*) breeding in suburban gardens. *Journal of Animal Ecology* 57:611–626.

Cowie, R. J., and S. A. Hinsley. 1988b. The provision of food and the use of bird feeders in suburban gardens. *Bird Study* 35:163–168.

Cox, D., and K. Gaston. 2016. Urban bird feeding: Connecting people with nature. *PLoS One* 11(7) doi: 10.1371/journal.pone.0158717.

Cuthill, I. C., and W. A. Macdonald. 1990. Experimental manipulation of the dawn and dusk chorus in the blackbird *Turdus merula*. *Behavioral Ecology and Sociobiology* 26:209–216.

Davies, Z. G., R. A. Fuller, M. Dallimer, A. Loram, and K. J. Gaston. 2012. Household factors influencing participation in bird feeding activity: A national scale analysis. *PLoS ONE* 7(6), e39692.

Davis, A., C. E. Taylor, and R. E. Major. 2011. Do fire and rainfall drive spatial and temporal population shifts in parrots? A case study using urban parrot populations. *Landscape and Urban Planning* 100:295–301.

Deis, R. 1982. Is bird feeding a no-no? *Defenders* 54:17–18.

Dhondt, A. A., S. Altizer, E. G. Cooch, A. K. Davis, A. Dobson, M. Driscoll, B. Hartup, D. Hawley, W. Hochachka, P. Hosseini, C. Jennelle, G. Kollias, D. S. Ley, C. Elliot, and K. Sydenstricker. 2005. Dynamics of a novel pathogen in an avian host: Mycoplasmal conjunctivitis in house finches. *Acta Tropica* 94:77–93.

Dhont, A. A., M. J. L. Driscoll, and E. C. H. Swarthout. 2007. House finch *Carpodacus mexicanus* roosting behavior during the non-breeding season and possible effects of mycoplasmal conjunctivitis. *Ibis* 149:1–9.

Dhondt, A. A., R. Eyckerman, R. Moermans, and R. Huble. 1984. Habitat and laying date of great and blue tit *Parus major* and *P. caeruleus*. *Ibis* 126:388–397.

Dhondt, A. A., D. L. Tessaglia, and R. L. Slothower. 1995. Epidemic mycoplasmal conjunctivitis in house finches from Eastern North America. *Journal of Wildlife Diseases* 34:265–280.

Drent, R. H., and S. Daan. 1980. The prudent parent: Energetic adjustments in avian breeding. *Ardea* 68: 225–252.

Dubois, S., and D. Fraser. 2013. A framework to evaluate wildlife feeding in research, wildlife management, tourism and recreation. *Animals* 3(4): 978–994.

Edmunds, K., N. Bunbury, S. Sawmy, C. G. Jones, and D. J. Bell. 2008. Restoring avian island endemics: Use of supplementary food by the endangered Pink Pigeon (*Columba mayeri*). *Emu* 108:74–80.

Fischer, J. D., and J. R. Miller. 2015. Direct and indirect effects of anthropogenic bird food on population dynamics of a songbird. *Acta Oecologica* 69:46–51.

Fischer, J. R., D. E. Stallknecht, M. P. Luttrell, A. A. Dhondt, and K. A. Converse. 1997. Mycoplasmal conjunctivitis in wild songbirds: The spread of a new contagious disease in a mobile host population. *Emerging Infectious Diseases* 3:69–72.

Flannery, T. F. 2005. *The Future Eaters: An Ecological History of the Australasian Lands and People*. Sydney: New Holland Publishers.

Fleischer, A. L., R. Bowman, and G. E. Woolfenden. 2003. Variation in foraging behavior, diet and time of breeding in Florida Scrub-Jays in suburban and wildland habitats. *Condor* 105:515–527.

Floyd, R. B., and D. J. Woodland. 1981. Localization of soil dwelling scarab larvae by the black-backed magpie, *Gymnorhina tibicen* (Latham). *Animal Behaviour* 29(2): 510–517.

Forbush, E. H. 1918. *Food, Feeding and Drinking Appliances and Nesting Material to Attract Birds*. Boston: Wright and Potter.

French, K., R. E. Major, and K. Hely. 2004. The role of nectar-producing plants in attracting birds to gardens. *Biological Conservation* 121:545–559.

Frith, C. B., and D. W. Frith. 2001. Nesting biology of the spotted catbird in Australian Wet Tropics upland rainforests. *Australian Journal of Zoology* 49:279–310.

Fuller, R. A., K. Irvine, Z. Davies, P. R. Armsworth, and K. J. Gaston. 2012. Interactions between people and birds in urban landscapes. *Urban Bird Ecology and Conservation* 45:249–266.

Fuller, R. A., and K. N. Irvine. 2010. Interactions between people and nature in urban environments. *In* K. J. Gaston, ed., *Urban Ecology*, pp. 134–171. Cambridge: Cambridge University Press.

Fuller, R. A., P. H. Warren, P. R. Armsworth, O. Barbosa, and K. J. Gaston. 2008. Garden bird feeding predicts the structure of urban avian assemblages. *Diversity and Distributions* 14:131–137.

Galbraith, J. A., J. R. Beggs, D. N. Jones, E. J. McNaughton, C. R. Krull, and M. C. Stanley. 2014. Risks and drivers of wild bird feeding in urban areas of New Zealand. *Biological Conservation* 180:64–74.

Galbraith, J. A., J. R. Beggs, D. N. Jones, and M. C. Stanley. 2015. Supplementary feeding restructures urban bird communities. *Proceedings of the National Academy of Sciences* 112(20): E648–657.

Ganas, P., B. Jaskulska, B. Lawson, M. Zadravec, M. Hess, and I. Bilic. 2014. Multilocus sequence typing confirms the clonality of *Trichomonas gallinae* isolates circulating in European finches. *Parasitology* 141:652–661.

Gaston, K., R. A. Fuller, A. Loram, C. MacDonald, S. Power, and N. Dempsey. 2007. Urban domestic gardens (XI): Variation in urban wildlife gardening in the UK. *Biodiversity and Conservation* 14:3227–3238.

Geis, A. D. 1980. *Relative Attractiveness of Different Foods at Wild Bird Feeders*. US Fish and Wildlife Service Special Scientific Report 117.

Glue, D. E. 1982. *The Garden Bird Book*. London: Macmillan.

Glue, D. 2006. Variety at winter bird tables. *Bird Populations* 7:212–215.

Greenaway, T. J. 1988. *Food Preference of Wild Birds: Seeds, Nuts and Composite Mixtures*. Tring, UK: Research Report 137, British Trust for Ornithology.

GrrlScientist. 2012. Scotts Miracle-Gro: The bird-killing company. *The Guardian*. http://www.theguardian.com/science/grrlscientist/2012/mar/21/2?newsfeed=true.

Grubb, T. C. 1987. Changes in the flocking behaviour of wintering English titmice with time, weather and supplementary food. *Animal Behaviour* 35:794–806.

Hanson, R. P. 1969. The possible role of infectious agents in the extinction of species. In J. J. Hickey, ed., *Peregrine Falcon Populations*, pp. 439–444. Madison: University of Wisconsin Press.

Harrison, T. J. E., J. A. Smith, G. R. Martin, D. E. Chamberlain, S. Bearhop, G. N. Robb, and S. J. Reynolds. 2010. Does food supplementation really enhance productivity of breeding birds? *Oecologia* 164(2): 311–320.

Hatchwell, B. J. (2009). The evolution of cooperative breeding in birds: Kinship, dispersal and life history. *Philosophical Transactions of the Royal Society B* 364:3217–3227.

Hiesemann, M. 1908. *How to Attract and Protect Wild Birds.* Translated by E. Buchheim. London: Witherby.

Hill, W. L. 1988. The effect of food abundance on the reproductive patterns on coots. *Condor* 90:324–331.

Hofle, U., C. Gortazar, J. A. Ortiz, B. Knispel, and E. F. Kaleta. 2004. Outbreak of trichomoniasis in a woodpigeon (*Columba palumbus*) wintering roost. *European Journal of Wildlife Research* 50:73–77.

Hosseini, P. R., A. A. Dhondt, and A. Dobson. 2004. Seasonality and wildlife diseases: How seasonal birth, aggregation and variation in immunity affect the dynamics of *Mycoplasma gallisepticum* in house finches. *Proceedings of the Royal Society of London B*, 271:2569–2577.

Houston, D., K. Mcinnes, G. Elliott, D. Eason, R. Moorhouse, and J. Cockrem. 2007. The use of a nutritional supplement to improve egg production in the endangered kakapo. *Biological Conservation* 138:248–255.

Howard, P. 2006. The beast within: An exploration of Australian constructions of wildlife. PhD thesis, Griffiith University, Brisbane, Australia.

Howard, P., and D. N. Jones. 2004. A qualitative study of wildlife feeding in south-east Queensland. *In* S. Burgin and D. Lunney, eds., *Urban Wildlife: More Than Meets the Eye,* pp. 55–62. Sydney: Royal Zoological Society of NSW.

Howkins, A., J. Orsi, and M. Fiege, eds. 2012. *National Parks beyond the Nation.* Norman: University of Oklahoma Press.

Hughes, J. M., P. B. Mather, A. Toon, J. Ma, I. Rowley, and E. Russell. 2003. High levels of extra-group paternity in a population of Australian magpies *Gynorhina tibicen*: Evidence from mircosatellite analysis. *Molecular Ecology* 12:3441–3450.

Ishigame, G., and G. S. Baxter. 2007. Practice and attitudes of suburban and rural dwellers to feeding wild birds in Southeast Queensland, Australia. *Ornithological Sciences* 6:11–19.

Ishigame, G., G. S. Baxter, and A. T. Lisle. 2006. Effects of artificial foods on the blood chemistry of the Australian magpie. *Austral Ecology* 31(2): 199–207.

Jaggard, A. K., N. Smith, F. R. Torpy, and U. Munro. 2015. Rules of the roost: Characteristics of nocturnal communal roosts of rainbow lorikeets *(Trichoglossus haematodus*, Psittacidae) in an urban environment. *Urban Ecosystems* 18:489–502.

James, K. A. C., G. C. Waghorn, R. G. Powlesland, and B. D. Lloyd. 1991. Supplementary feeding of Kakapo on Little Barrier Island. *Proceedings of the Nutritional Society of New Zealand* 16:93–102.

Jansson, C., J. Ekman, and A. Bromssen. 1981. Winter mortality and food supply in tits *Parus* spp. *Oikos* 37:313–322.

Jones, D. N. 1981. Temporal changes in the suburban avifauna of an inland city. *Australian Wildlife Research* 8:109–119.

Jones, D. N. 1983. The suburban bird community of Townsville, a tropical city. *Emu* 83(1): 12–18.

Jones, D. N. 2002. *Magpie Alert: Learning to Live with a Wild Neighbour.* Kensington: New South Wales University Press.

Jones, D. 2011. An appetite for connection: Why we need to understand the effect and value of feeding wild birds. *Emu* 111:i–vii.

Jones, D. 2014. It's time to talk about feeding. *Australian Birdlife* 3(2): 8.

Jones, D., and P. Howard. 2001. Feeding wildlife in urban areas: An indecent obsession? *Wildlife Australia* 38(3): 18–20.

Jones, D. N., and S. James Reynolds. 2008. Feeding birds in our towns and cities: A global research opportunity. *Journal of Avian Biology* 39:265–271.

Kacelnik, A. 1979. The foraging efficiency of great tits (*Parus major* L.) in relation to light intensity. *Animal Behaviour* 27:237–241.

Kallander, H. 1981. The effects of provision of food in winter on a population of the Great Tit *Parus major* and the Blue Tit *P. caeruleus*. *Ornis Scandinavica* 12:244–248.

Kellert, S. R. 1996. *The Value of Life: Biological Diversity and Human Society*. Washington, DC: Island Press.

Kellert, S. R. 1997. *Kinship to Mastery: Biophilia in Human Evolution and Development*. Washington, DC: Island Press.

Kluyver, H. N. 1952. Notes on body weight and time of breeding in the Great Tit, *Parus m. major* L. *Ardea* 40:123–141.

Kobilinsky, D. 2015. Supplementary feeding: To feed or not to feed wildlife. The Wildlife Society website, April 8, 2015. tws.wildlife.org.

Kubota, H., and M. Nakamura. 2000. Effects of supplemental food on intra and interspecific behaviour of the Varied Tit *Parus varius*. *Ibis* 142:312–319.

Lawson, B., S. MacDonald, T. Howard, S. K. Macgregor, and A. A. Cunningham. 2006. Exposure of garden birds to aflatoxins in Britain. *Science of the Total Environment* 361(1–3): 124–131.

Lawson, B., R. A. Robinson, K. M. Colvile, K. M. Peck, J. Chantrey, T. W. Pennycott, V. R. Simpson, M. P. Toms, and A. A. Cunningham. 2012. The emergence and spread of finch trichomonosis in the British Isles. *Philosophical Transactions of the Royal Society B* 367:2852–2863.

Lepczyk, C. A., A. G. Mertig, and J. Liu. 2004. Assessing landowner activities related to birds across rural-to-urban landscapes. *Environmental Management* 33(1):110–125.

Lepczyk, C. A., P. S. Warren, L. Machabee, A. P. Kinzig, and A. G. Mertig. 2012. Who feeds the birds? A comparison across regions. *Urban Bird Ecology and Conservation* 45:267–284.

Leston, L. F., and A. D. Rodewald. 2006. Are urban forests ecological traps for understory birds? An examination using Northern cardinals. *Biological Conservation* 131:566–574.

Leung, M. C. K., G. Díaz-Llano, and T. K. Smith. 2006. Mycotoxins in pet food: A review on worldwide prevalence and preventative strategies. *Journal of Agricultural and Food Chemistry* 54(26): 9623–9635.

Ley, D. H., J. E. Berkhoff, and J. M. McLaren. 1996. Mycoplasma gallisepticum isolated from house finches (*Carpodacus mexicanus*). *Avian Diseases* 40:480–483.

Lin, E. 2005. *Production and Processing of Small Seeds for Birds*. Rome: United Nations Food and Agriculture Organisation.

Lloyd, B. D., and R. G. Powlesland. 1994. The decline of kakapo *Strigops habroptilus* and attempts at conservation by translocation. *Biological Conservation* 69:75–85.

Louv, R. 2005. *Last Child in the Woods*. New York: Workman Publishing.

Low, T. 2002. *The New Nature*. Melbourne: Viking.

Low, T. 2014. *Where Song Began*. Sydney: Penguin Australia.

Manfredo, M. J. 2008. *Who Cares about Wildlife? Social Science Concepts for Exploring Human-Wildlife Relationships and Conservation Issues*. New York: Springer.

Martin, T. E. 1987. Food as a limit on breeding birds: A life-history perspective. *Annual Review of Ecology and Systematics* 18:454–487.

Martin, T. E., P. R. Martin, C. R. Olson, B. J. Heidinger, and J. J. Fontaine. 2000. *Parental Care and Clutch Sizes in North and South American Birds*. Lincoln: Nebraska Cooperative Fish and Wildlife Research Unit.

Marzluff, J. M. 2014. *Welcome to Subirdia: Sharing Our Neighborhoods with Wrens, Robins, Woodpeckers, and Other Wildlife*. New Haven, CT: Yale University Press.

Marzluff, J. M., and T. Angell. 2005. *In the Company of Crows and Ravens*. New Haven, CT: Yale University Press.

Marzluff, J. M., and T. Angell. 2012. *Gifts of the Crow*. New York: Free Press.

Marzluff, J. M., R. Bowman, and R. Donnelly, eds. 2001. *Avian Ecology and Conservation in an Urbanizing World*. Boston: Kluwer Academic.

Marzluff. J. M., and M. Miller. 2014. Crows and crow feeders: Observations on interspecific semiotics. *In* G. Witzany, ed., *Biocommunication of Animals*, pp. 191–211. Dordrecht: Springer Press.

McAtee, W. L. 1914. *How to Attract Birds in Northeastern United States*. Washington, DC: U.S. Department of Agriculture.

McCarthy, B. M. 2006. Shakespeare's red kite returns to London after an absence of 150 years. *The Independent*, 13 January 2006.

McCormick, I., C. W. Davison, and R. L. Hoskin. 1992. *The U.S. Sunflower Industry*. Herndon, VA: US Department of Agriculture.

McDonnell, M. J., A. K. Hahs, and J. H. Breuste, eds. 2009. *Ecology of Cities and Towns: A Comparative Approach*. Cambridge: Cambridge University Press.

McLees, B. 2001. Feeding wildlife: Right or wrong? Community attitudes towards feeding wildlife in Melbourne, Australia and implications for management. BSc thesis, Deakin University, Melbourne, Australia.

Meijer, T., and R. Drent. 1999. Re-examination of the capital and income dichotomy in breeding birds. *Ibis* 141:399–414.

Miller, J. R. 2005. Biodiversity conservation and the extinction of experience. *Trends in Ecology and Evolution* 20:430–434.

Nattrass, R. 2001. To feed or not to feed. *Wildlife Australia* 38(3): 22.

O'Leary, R., and D. Jones. 2006. The use of supplementary foods by Australian magpies *Gymnorhina tibicen*: Implications for wildlife feeding in suburban environments. *Austral Ecology* 31:208–216.

Orros, M. E., and M. D. E. Fellowes. 2014. Supplementary feeding of the reintroduced Red Kite *Milvus milvus* in UK gardens. *Bird Study* 61(2).

Orros, M. E., and M. D. E. Fellowes. 2015. Wild bird feeding in an urban area: Intensity, economics and numbers of individuals supported. *Acta Ornithologica* 50:43–58.

Otter, K., B. Chruszcz, and L. Ratcliffe. 1997. Honest advertisement and song output during the dawn chorus of black-capped chickadees. *Behavioral Ecology* 8:167–173.

Ottoni, I., F. F. R. De Oliveira, and R. J. Young. 2009. Estimating the diet of urban birds: The problems of anthropogenic food and food digestibility. *Applied Animal Behaviour Science* 117:42–46.

Parra, J., and J. L. Tellería. 2004. The increase in the Spanish population of Griffon Vulture *Gyps fulvus* during 1989–1999: Effects of food and nest site availability. *Bird Conservation International* 14:33–41.

Parsons, H. 2008. Guidelines for enhancing urban bird habitat *Wingspan* 18(1): 24–27.

Parsons, H., R. E. Major, and K. French. (2006). Species interactions and habitat associations of birds inhabiting urban areas of Sydney, Australia. *Austral Ecology* 31:217–227.

Pawson, E., and T. Brooking. 2002. *Environmental Histories of New Zealand*. Oxford: Oxford University Press.

Peach, W. J., J. W. Mallord, N. Ockendon, C. J. Orsman, and W. G. Haines. 2015. Invertebrate prey availability limits reproductive success but not breeding population size in suburban House Sparrows *Passer domesticus. Ibis* 157:601–613.

Peach, W. J., D. Sheehan, and W. Kirby. 2014. Supplementary feeding of mealworms enhances reproductive success in garden nesting House Sparrows *Passer domesticus. Bird Study* 61:378–385.

Peach, W. J., K. E. Vincent, J. A. Fowler, and P. V. Grice. 2008. Reproductive success of house sparrows along an urban gradient. *Animal Conservation* 11:493–503.

Peck, H. L., H. E. Pringle, H. H. Marshall, I. P. F. Owens, and A. M. Lord. 2014. Experimental evidence of impacts of an invasive parakeet on foraging behavior of native birds. *Behavioral Ecology* 25:582–590.

Perkins, H. E. 2010. Measuring love and care for nature. *Journal of Environmental Psychology* 30:455–463.

Peterson, R. T. 2000. *Feeder Birds: Eastern North America*. Boston: Houghton Mifflin.

Plant, M. 2008. Good practice when feeding birds. *Wingspan* 18(1): 20–23.

Plummer, K. E., S. Bearhop, D. I. Leech, D. E. Chamberlain, and D. Blount. 2013. Fat provisioning in winter impairs egg production during the following spring: A landscape-scale study of blue tits. *Journal of Animal Ecology* 82:673–682.

Plummer, K. E., G. M. Siriwardena, G. J. Conway, K. Risely, and M. P. Toms. 2015. Is supplementary feeding in gardens a driver of evolutionary change in a migratory bird species? *Global Change Biology* 21:353–363.

Poesel, A., H. P. Kunc, K. Foerster, A. Johnsen, and B. Kempenaers. 2006. Early birds are sexy: Male age, dawn song and extrapair paternity in blue tits, *Cyanistes* (formerly *Parus*) *caeruleus. Animal Behaviour* 72:531–538.

Powlesland, R. G., and B. D. Lloyd. 1994. Use of supplementary feeding to induce breeding in free-living kakapo *Strigops habroptilus* in New Zealand. *Biological Conservation* 69:97–106.

Reynolds, S. J., S. J. Schoech, and R. Bowman. 2003. Nutritional quality of prebreeding diet influences breeding performance of the Florida scrub-jay. *Oecologia* 134:308–316.

Robb, G. N., R. A. McDonald, D. E. Chamberlain, and S. Bearhop. 2008a. Food for thought: Supplementary feeding as a driver of ecological change in avian populations. *Frontiers in Ecology and the Environment* 6:476–484.

Robb, G. N., R. A. McDonald, D. E. Chamberlain, S. J. Reynolds, T. J. Harrison, and S. Bearhop. 2008b. Winter feeding of birds increases productivity in the subsequent bredding season. *Biology Letters* 4:220–223.

Robbins, C. T. 1983. *Wildlife Feeding and Nutrition*. San Diego: Academic Press.

Robinson, R. A., B. Lawson, M. P. Toms, K. M. Peck, J. K. Kirkwood, J. Chantrey, I. Clatworthy, A. Wvans, L. Hughes, O. Hutchinson, S. John, T. Pennycott, M. Perkins, P. Rowley, V. Simpson, K. Tyler, and A. A. Cunningham. 2010. Emerging infectious diseases leads to rapid population declines of common British birds. *PLoS ONE* 5(8): e12215.

Rollinson, D. D. J., and D. N. Jones. 2002. Variation in breeding parameters of the Australian magpie *Gymnorhina tibicen* in suburban and rural environments. *Urban Ecosystems* 6:257–269.

Rollinson, D. J., R. O'Leary, and D. N. Jones. 2003. The practice of wildlife feeding in suburban Brisbane. *Corella* 27:52–58.

Rolls, E. 1969. *They All Ran Wild: The Animals and Plants That Plague Australia*. Sydney: Angus and Robertson.

Rolshausen, G., G. Segelbacher, K. A. Hobson, and H. M. Schaefer. 2009. Contemporary evolution of reproductive isolation and phenotypic divergence in sympatry along a migratory divide. *Current Biology* 19:2097–2101.

Saggese, K., F. Korner-Nievergelt, T. Slagsvold, and V. Amrhein. (2011). Wild bird feeding delays start of dawn singing in the great tit. *Animal Behaviour* 81(2): 361–365.

Schoech, S. J. 1996. The effect of supplementary food on body condition and the timing of reproduction in a cooperative breeder, the Florida scrub-jay. *Condor* 98:234–244.

Schoech, S. J., and R. Bowman. 2003. Does differential access to protein influence differences in timing of breeding of Florida scrub-jays (*Aphelocoma coerulescens*) in suburban and wildland habitats? *Auk* 120:1114–1127.

Schreiber, L. A. 2010. Why we feed wild birds: A case study of BTO members' motivation for feeding birds in their gardens. MSc thesis, University College, London.

Selhub, E. M., and A. C. Logan. 2012. *Your Brain on Nature: The Science of Nature's Influence on Your Health, Happiness and Vitality*. Mississauga: John Wiley & Sons Canada.

Smith, J., and A. Lill. 2008. Importance of eucalypts in exploitation of urban parks by Rainbow and Musk Lorikeets. *Emu* 108:187–195.

Smith, S. M. 1967. Seasonal changes in the survival of the black-capped chickadee. *Condor* 69:344–359.

Smith, S. M. 1991. *The Black-Capped Chickadee: Behavioral Ecology and Natural History*. Ithaca, NY: Comstock.

Soper, T. 1965. *The Bird Table Book*. Newton Abbott: David & Charles.

South, J., and S. Pruett-Jones. 2000. Patterns of flock size, diet, and vigilance of naturalized monk parakeets in Hyde Park, Chicago. *Condor* 102:848–854.

Sterba, J. P. 2002. American backyard feeders may do harm to wild birds. *Wall Street Journal*, December 27.

Sterba, J. P. 2012. *Nature Wars: The Incredible Story of How Wildlife Comebacks Turned Backyards into Battlegrounds*. New York: Crown Publishers.

Stolzenburg, W. 2011. *Rat Island: Predators in Paradise and the World's Greatest Wildlife Rescue*. London: Bloomsbury.

Summers-Smith, J. D. 2003. The decline of the house sparrow: A review. *British Birds* 96:439–446.

Svensson, E., and J.-A. Nilsson. 1995. Food supply, territory quality and reproductive timing in the blue tit (*Parus caeruleus*). *Ecology* 76:1804–1812.

Swainson, C. (1885) 2004. *The Folk Lore and Provincial Names of British Birds*. London: Kessinger.

Taylor, S., and I. Castro. (2001). *Standard Operating Procedure Manual for Hihi*. Wellington: Department of Conservation, New Zealand.

Tella, J. L. 2001. Sex-ratio theory in conservation biology. *Trends in Ecology and Evolution* 16:76–77.

Tennyson, A., and P. Martinson. 2006. *Extinct Birds of New Zealand*. Christchurch, New Zealand: Te Papa Press.

Thomas, L. 2000. Wildlife and humans in a suburban setting: Understanding wildlife-human interactions in South-East Queensland. PhD thesis, Griffith University, Brisbane, Australia.

Thomas, N. J., D. B. Hunter, and C. T. Atkinson, eds. 2007. *Infectious Diseases in Wild Birds*. Melbourne: Blackwell Publishing.

Thompson, A. 2012. *Francis of Assisi: A New Biography*. Ithaca, NY: Cornell University Press.

Thompson, P. S. 1987. *The seasonal use of gardens by birds with special reference to supplementary feeding*. Tring, UK: Research Report 27, British Trust for Ornithology.

Thompson, R. 2011. *The Plume Hunter*. Torrey, UT: Torrey House Press.

Thoreau, H. D. (1854) 1946. *Walden; or, Life in the Woods*. New York: Dodd, Mead.

Todd, K. 2012. *Sparrow*. London: Reaktion Books.

Tollington, S., A. Greenwood, C. G. Jones, P. Hoeck, D. Smith, H. Richards, and V. Tatayah. 2015. Detailed monitoring of a small but recovering population reveals sublethal effects of disease and unexpected interactions with supplemental feeding. *Journal of Animal Ecology* 84:969–977.

Toms, M., and P. Sterry. 2008. *Garden Birds and Wildlife*. Tring, UK: British Trust for Ornithology/AA Publishing.

US Fish and Wildlife Service. 2012. *National Survey of Fishing, Hunting and Wildlife-Associated Recreation*. Arlington, VA: US Fish and Wildlife Service.

Vickery, J. A., R. B. Bradbury, I. G. Henderson, M. A. Eaton, and P. V. Grice. 2004. The role of agri-environmental schemes and farm management practices in reversing the decline of farmland birds in England. *Biological Conservation* 119:19–39.

von Berlepsch, H. F. 1899. *Der gesamte Vogelschutz: Seine Bergundung und Ausfuhrung*. Kohler: Gera-Untermhaus.

Wade, A. D., S. Ikram, G. Conlogue, R. Beckett, A. J. Nelson, R. Colten, B. Lawson, and D. Tampieri. 2012. Foodstuff placement in ibis mummies and the role of viscera in embalming. *Journal of Archaeological Science* 39:1642–1647.

Waters, M. N., M. F. Piehler, J. M. Smoak, and C. S. Martens. 2009. The development and persistence of alternative ecosystem states in a large, shallow lake. *Freshwater Biology* 55:1249–1261.

Watson, A., ed. 1970. *Animal Populations in Relation to Their Food Resources.* Oxford: Blackwell Scientific Publications.

Wilcoxen, T., D. Horn, B. Hogan, C. Hubble, S. Huber, J. Flamm, M. Knott, L. Lundstrom, F. Salik, S. Wassenhove, and E. Wrobel. 2015. Effects of bird-feeding activities on the health of wild birds. *Conservation Physiology*, 3. doi: 10.1093/conphys/cov058.

Wilson, K.-J. 2004. *Flight of the Huia.* Christchurch, New Zealand: Canterbury University Press.

Wilson, W. H. 2001. The effects of supplementary feeding on Black-capped Chickadees (*Poecile atricapilla*) in Central Maine: Population and individual responses. *Wilson Bulletin* 113:65–72.

Woolfenden, G. E., and J. W. Fitzpatrick. 1984. *The Florida Scrub Jay: Demography of a Cooperative-Breeding Bird.* Princeton, NJ: Princeton University Press.

Worthy, T. H., and R. N. Holdaway. 2002. *The Lost World of the Moa: Prehistoric Life of New Zealand.* Christchurch, New Zealand: Canterbury University Press.

Wright, J., and M. Leonard, eds. 2007. *The Evolution of Begging: Competition, Cooperation and Communication.* New York: Kluwer Academic.

Index